D1064610

Results and Problems in Cell Differentiation

A Series of Topical Volumes in Developmental Biology

6

Editors

W. Beermann, Tübingen · J. Reinert, Berlin · H. Ursprung, Zürich

W. J. Dickinson · D. T. Sullivan

Gene-Enzyme Systems in Drosophila

With 32 Figures

Springer-Verlag New York Heidelberg Berlin 1975

Dr. WILLIAM J. DICKINSON
Assistant Professor of Biology
Department of Biology
The University of Utah
Salt Lake City, Utah 84112, USA

Dr. DAVID T. SULLIVAN
Associate Professor of Biology
Department of Biology
Syracuse University
Syracuse, N.Y. 13210, USA

ISBN 0-387-06977-1 Springer-Verlag New York Heidelberg Berlin
ISBN 3-540-06977-1 Springer-Verlag Berlin Heidelberg New York

Library of Congress Cataloging in Publication Data.
Dickinson, William J., 1940–. Gene-enzyme systems in drosophila. (Results and problems in cell differentiation, 6). Bibliography: p. 1. Drosophila. 2. Biochemical genetics. 3. Enzymes. 4. Insect genetics. I. Sullivan, David T., 1940– joint author. II. Title. III. Series. [DNLM: 1. Drosophila-Enzymology. 2. Eye. 3. Genes. 4. Mutation. W1RE248X v. 6/QH470.D7 D553g]. QH607.R4 vol. 6 [QH470.D7] 574.8'761s [595.7'74]. 74–17430.

Typesetting, printing and bookbinding: Brühlsche Universitätsdruckerei, Gießen.

Preface

There was a period in the history of modern biology when proteins were thought to be "gene products" in a rather direct sense. An account of their appearance and disappearance in the course of development and differentiation thus seemed an appropriate means of describing "gene regulation".

When RNA was found to be the immediate product of genetic activity, the study of proteins as gene products lost some of its original attraction. Indeed, the development of the powerful method of nucleic acid hybridization aroused the hope that a large array of specific messenger-RNA molecules synthesized during cell differentiation could be individually assayed. The difficulties in the way of such ambitious projects were described in Volume 3 of this series: *Nucleic Acid Hybridization in the Study of Cell Differentiation* (ed. H. URSPRUNG, 1972).

Enzymes are in large measure responsible for cell function. Clearly, their synthesis must be under genetic control. We are convinced that the study of enzyme behavior during development merits much attention, particularly if the work is carried out on a eukaryote that lends itself to genetic and developmental analysis. An impressive amount of genetic information is available on the insect *Drosophila*, and much has been learnt about its development. The giant chromosomes present in several tissues of this insect might well be useful in a continuing analysis of the appearance of specific enzymes and the transcription of the cognate genetic loci.

The present volume is an up-to-date account of gene-enzyme systems in *Drosophila*. Written for biochemists, molecular biologists, and cell biologists, it provides access to a large body of literature on this subject.

Tübingen, Berlin, Zürich
October 1974

W. BEERMANN
J. REINERT
H. URSPRUNG

Contents

Genetic Loci Cited

Full name	Abbreviation
Acid phosphatase	Acph
Alcohol dehydrogenase	Adh
Aldehyde oxidase	Aldox
Amylase	Amy
Alkaline phosphatase	Aph
black	b
Blond	Bld
brown	bw
blistery	by
curved	c
claret	ca
cardinal	cd
cinnamon	cin
cinnabar	cn
Dichaete	D
double sex	dsx
ebony	e
elbow	el
Esterase	Est
garnet	g
Glued	Gl
α-glycerophosphate dehydrogenase	α-Gpdh
Hexokinase	Hex
Isocitrate dehydrogenase	Idh
javelin	jv
lozenge	lz
low aldehyde oxidase	lao
Leucine amino peptidase	Lap
low isoxanthopterine	lix
light	lt
lightoid	ltd
lethal meander	lme
low pyridoxal oxidase	lpo
low xanthine dehydrogenase	lxd
maroon	ma
maroon-like	ma-l
Minute	M
Octanol dehydrogenase	Odh
pink	p
Phosphogluconate dehydrogenase	Pgd
Phosphoglucomutase	Pgm
prune	pn
purple	pr
reduced-scraggly	rd^s
rose	rs
roughoid	ru
rosy	ry
stubble	sb
sepia	se
sepiaoid	sed
Serate	Ser
spineless	ss
scarlet	st
straw	stw
supressor of sable	su-(s)
thread	th
transformer	tra
vermilion	v
vestigal	vg
white	w
white-apricot	w^a
welt	wt
yellow	y
Zwischenferment (glucose-6-phosphate dehydrogenase)	Zw

The Study of Gene-Enzyme Systems

Introduction

The decision to write a book on gene-enzyme systems in *Drosophila* grew out of a conviction that systems which are well-defined both genetically and biochemically have been and will continue to be useful in approaching a wide variety of problems, particularly relative to the organization, function and regulation of the eukaryotic genome. At the same time, our own experience indicated that the literature on such well-defined gene-enzyme systems is widely scattered and not easily accessible. We felt it would be useful to bring together and summarize a substantial proportion of this work. The present book is the result.

Interest in gene-enzyme systems in *Drosophila* is not entirely new. Early work on the biochemical basis of eye color mutations in *Drosophila,* notably that by BEADLE *et al.* during the 1930's, played a crucial role in establishing the relationship between genes and enzymes (see p. 32). In the following two or three decades, most workers interested in the biochemical aspects of genetics turned to less complex organisms with, of course, enormous success. During the last ten to fifteen years, there has been a considerable renewal of interest in gene-enzyme systems in *Drosophila.* This revival is attributable in part to the development of methods that facilitate the work, and in part to interest in phenomena unique to eukaryotes or the desire to test the generality of models developed on the basis of work with prokaryotes. Given the wealth of information generated by hundreds of investigators over more than sixty years, and the sophistication of the genetic methods possible with this organism, *Drosophila* clearly is a logical choice for this kind of work.

In writing this book, we had three objectives in mind:

1. To provide a reasonably complete description of each gene-enzyme system that has been studied to date;

2. to give an introduction to the methods most useful for work on gene-enzyme systems;

3. to indicate the range of problems which may be attacked using gene-enzyme systems.

The subsequent chapters contain the descriptions of the various gene-enzyme systems, grouped where possible according to genetic or biochemical relationships. Since one of the principal advantages of these systems is the possibility of bringing genetic, biochemical and developmental information to bear on the same problem, all of the available information on a given enzyme has been kept together. The remainder of this chapter serves as survey of the methods and applications, and as a guide to where detailed examples of these methods and applications can be found in the subsequent chapters.

Applications

Most of the recent work on *Drosophila* gene-enzyme systems has been prompted by interest in one or more of the following four areas:

1. The regulation of gene function in eukaryotes, particularly as it relates to development;
2. the organization and fine structure of the eukaryotic genome;
3. aspects of population genetics and evolution;
4. biochemistry, physiology and metabolic pathways.

In accordance with the title of the series in which this volume is published, as well as our own interests and competence, we will emphasize work relevant to the first two areas. However, we have drawn upon publications in which the primary interest is in one of the other areas when this contributes to the basic description of any gene-enzyme system.

Development

Perhaps the most promising and exciting applications of gene-enzyme systems are in developmental biology. They can serve as model systems for studying gene regulation, and as markers for tracing a variety of developmental events.

Models of Differential Gene Function

Attempts to use genetic methods to analyze gene function during development are by no means new. Much of the classical work in developmental genetics used mutations which are lethal, or which have dramatic morphological effects. Such mutations were analyzed to determine the lethal phase, the array (often complex) of phenotypic effects, the time of earliest indication of abnormality, autonomy of expression in various tissues, etc. (see WRIGHT, 1970). The use of temperature-sensitive lethal mutations (SUZUKI, 1970) is essentially a refinement of this approach. All of this work has in common a very basic shortcoming. Any analysis, no matter how detailed or careful, can, at best, determine when and possibly where the gene product performs an essential function. This approach can never lead to direct evidence on when and where the gene product is made. Information on this last point can be obtained only if the normal product of the locus being studied can be identified and measured directly. Identification of the primary biochemical defect underlying a morphological phenotype has been extremely difficult, and has been accomplished in only a handful of cases among the hundreds of known mutants in *Drosophila*.

A particularly attractive alternative is to select a gene with a known product and attempt to understand the events and mechanisms involved in the production of that product. Genes with known enzyme products are obvious candidates. Focusing attention on a single enzyme provides a much needed simplification and permits quantitative measurements in place of qualitative descriptions. The benefits of concentrating on relatively simple, biochemically defined model systems have been amply demonstrated by the spectacular progress in understanding of the regulation of such systems in prokaryotes. This work also illustrates the usefulness of genetic tools in such analyses.

The choice of a gene-enzyme system for use in this kind of study might be influenced by a number of factors, including biochemical convenience, specificity of expression during development, availability of useful or interesting mutants or methods for obtaining mutants, and physiological significance.

Biochemical Convenience

Most of the enzymes included in this book have been studied in part because convenient assays exist. The dehydrogenases (Chapter 4), esterases and phosphatases (Chapter 7), have been popular subjects because of the easy methods for detecting enzyme activity in electrophoretic gels. For a multi-faceted study of a gene-enzyme system, availability of sensitive and specific methods for quantitative assay, gel staining and histochemical localization is desirable. Complex systems with multiple electrophoretic bands of enzyme activity, like those found with esterases and phosphatases, present problems for quantitative and histochemical methods, since unrelated enzymes overlap in substrate specificity. Particularly favorable systems in *Drosophila* include alcohol dehydrogenase (p. 71), aldehyde oxidase (p. 52), amylase ,(p. 107), and probably some of the other dehydrogenases (Chapter 4).

Developmental Specificity

To be useful in a study on gene regulation, an enzyme must be differentially expressed with respect to either timing or localization. Most of the enzymes that have been examined in this way do in fact show stage specific variations in total activity and/or tissue specific distribution of the activity. Stage specific variations in the activities of a series of enzymes are shown in Figs. 11, 22–26, and 29. Descriptive information about changes during development has been included in many other sections on individual enzymes. Particularly well studied cases include alcohol dehydrogenase (p. 76), aldehyde oxidase (p. 64), and alkaline phosphatase (p. 124). Dramatic evidence of changing patterns of gene function is furnished by a number of the multi-banded systems, in which various electrophoretic forms appear and disappear or change in relative concentration as development proceeds. Good examples include the isocitrate dehydrogenases (p. 93), hexokinases (p. 133), deoxyribonucleases (p. 136), esterases (p. 121), alkaline phosphatases (p. 124), and peptidases (p. 115). Although these changing patterns are usually attributed to changes in the relative activity of different genes, this interpretation must be made with caution. In at least one case (p. 126), it seems clear that a change actually reflects a secondary modification of an enzyme.

Evidence for tissue specificity of enzyme activity and/or electrophoretic banding pattern has been obtained by dissection and assay of separated tissues or organs in a number of cases. Among the enzymes studied in this way are kynurenine hydroxylase (p. 42), alcohol dehydrogenase (p. 77), isocitrate dehydrogenase (p. 93), hexokinase (p. 134), deoxyribonuclease (p. 137), xanthine dehydrogenase (p. 67), aldehyde oxidase (p. 65), esterases (p. 121), acid phosphatases (p. 129), alkaline phosphatases (p. 112), and amylases (p. 125). Histochemical staining of enzymes in sections or dissected organs can provide a more precise picture of enzyme distribution within organs or tissues, down to the cellular or even intra-cellular level. This kind of work has been done on aldehyde oxidase (p. 66). α-glycerophosphate dehydrogenase (p. 89), lactate dehydrogenase (p. 91), and alkaline phosphatose (p. 124). The results of this approach are often very dramatic (see Figs. 18 and 19).

Some special cases of gene regulation might also furnish useful model systems. The mechanism of dosage compensation for X-linked genes has been investigated in this context. Here again, X-linked genes with known enzyme products are most useful, since direct quantitative measurements are possible. Although the mechanism is not clear as yet, it is evident that compensation in *Drosophila* does not involve inactivation of one X-chromosome in each cell as is found in mammals. Some of the results are given in the sections on tryptophan oxygenase (p. 39), glucose-6-phosphate dehydrogenase (p. 82), and 6-phosphogluconate dehydrogenase (p. 85). A more complete review has been given by LUCHESSI (1973).

Sex specific differences in enzyme expression not related to sex linkage have also been reported. Some of these involve enzymes found in sex specific tissues like testes or paragonia (male accessory glands). In other cases, the significance of the sex specificity is less clear. Examples of sex limited enzymes include glucose-6-phosphate dehydrogenase (p. 83), hexokinase (p. 134), protease (p. 116), and some esterases (p. 121).

In the investigations of both sex linked and sex limited enzymes, the ready availability of flies with abnormal chromosome complements and the existence of mutations that influence the phenotypic expression of sex differences (e.g., *transformer* and *double sex*) provide useful tools. For example, some of the male specific enzymes were shown not to be Y-linked, since $\overline{XX}$/Y females do not make them and X0 males do. This again illustrates the advantage of working with an organism suitable for genetic manipulation.

Another aspect of the relationship between gene dosage and amount of gene product may have important implications for our understanding of gene regulation. In all cases where it has been possible to vary the number of copies of an enzyme structural gene, the level of enzyme activity is proportinal to the gene dosage (see p. 20). This clearly implies that the regulatory mechanism in these cases does not involve a feedback system to monitor and control the level of enzyme activity.

Mutations Which Modify Gene Expression

Mutations which alter the normal patterns of gene regulation have played a key role in the impressive advances in the understanding of regulation in prokaryotes. There is considerable hope that analogous mutations will be similarly useful in gaining some understanding of the differential gene expression central to eukaryotic development. The mutations which will be most interesting in this context are not those which alter the primary structure of a protein product, but rather those which modify the normal temporal or spatial pattern of expression, or which alter the responses to important environmental cues. Such mutations should serve to identify regulatory loci and permit some interesting questions about them to be asked. One would like to know how many regulatory loci influence a given gene, what range of effects they display, how they are organized in the genome relative to each other and the structural gene they affect, and how they interact with each other. Definitive evidence that a given mutation is regulatory rather than structural will only be possible if the gene product affected is biochemically well characterized, and the structural gene is known. Hence, well defined gene-enzyme systems once again are particularly favorable.

Few enzymes in *Drosophila* are known to be responsive to specific stimuli. Amylase activity is "induced" by starch in the diet (p. 112), and there is some evidence that xanthine dehydrogenase activity is modified under various culture conditions (p. 68). Presumably some of the changes stimulated by the various hormones that control insect development involve specific enzymes. At this writing, there are no reported cases of mutations which affect such responses.

A few cases of strain specific differences in developmental expression have been reported (see p. 8). Systematic searches for mutations causing such differences have begun recently (DICKINSON, 1971, 1972, 1974), and the results are promising. These few cases have depended on naturally occurring variability. There is some hope that strong selection methods, such as that devised for alcohol dehydrogenase (p. 76), will make possible strategies to recover some classes of regulatory mutants.

Cytogenetic Correlations

While gene-enzyme systems are partially defined at the biochemical level, they do leave open several steps (and possible regulatory mechanisms) between the gene and the finished enzyme product. Several years ago, URSPRUNG *et al.* (1968) proposed that this problem might be avoided for enzymes with known cytogenetic loci by studying puffing activity at the enzyme locus. The fine work on amylase (p. 112) seems to have established the merit of this approach.

Enzyme Mutants as Markers

Genetic markers have been used to "tag" specific cells for some time. Successful application of this approach requires a mutant phenotype that is recognizable at the single cell level and that is autonomous in expression, i.e., each cell has a phenotype characteristic of its own genotype. The classical markers used were pigment and bristle mutants readable only in cells involved in cuticle formation. A few enzyme mutants have been shown to be useful markers for internal tissues. Suitable histochemical methods permit recognition of the phenotype at the cell or tissue level. Autonomy of synthesis in specific tissues has been demonstrated by transplantation experiments in the cases of alcohol dehydrogenase (p. 77) and aldehyde oxidase (p. 68). Autonomy has been found in genetic mosaics for aldehyde oxidase (p. 68) and acid phosphatase (BENZER, 1973).

Once autonomy is established, enzyme markers may be used to investigate various problems concerning developmental fates and cell lineages. Marked cells may be introduced by transplantation, or may be induced by any of several procedures that produce genetic mosaics. POSTLETHWAIT'S and SCHNEIDERMAN'S (1973) review on developmental genetics of imaginal discs includes a survey of these methods. Representative of the problems that can be approached in this way is the demonstration by URSPRUNG *et al.* (1972) that at least part of the leg musculature is derived from imaginal disc cells. Muscle cells in the leg derived from an implanted leg disc contained enzyme markers present in the donor, but not the host. The fine method of producing fate maps for surface features using genetic mosaics recently elaborated by GARCIA-BELLIDO and MERIAM (1969) and HOTTA and BENZER (1972) has recently been extended to internal structures using enzyme markers (BENZER, 1973; JANNING, 1974a, 1974b).

Besides serving as cell and tissue markers, enzyme mutations may be used as indicators of gene activity. For example, the time of initial synthesis of an enzyme under control of the embryos' own genome can be estimated by determining the earliest appearance of a variant inherited from the paternal parent (WRIGHT and SHAW, 1970; see also pp. 65 and 88). In a slightly different context, enzyme mutations may be used to fill in the genetic map and serve as mapping marker genes for mutants which are expressed in earlier stages, but cannot be identified in adults. One example is the use of an esterase marker to map a larval alkaline phosphatase mutant (p. 123).

Drosophila as a Developmental System

The appeal of *Drosophila* gene-enzyme systems as models and tools in studying developmental problems is greatly enhanced by the virtual explosion of work on various aspects of development in *Drosophila*. The still useful general accounts of development in DEMEREC (1950) have been supplemented by the volumes edited by NOVITSKI and ASHBURNER (in press) and COUNCE and WADDINGTON (1973). Recent reviews on *Drosophila* development have covered development in general (FRISTROM, 1970), chromosome puffing (ASHBURNER, 1970), developmental genetics (WRIGHT, 1970), and imaginal discs (POSTLETHWAIT and SCHNEIDERMAN, 1973). URSPRUNG (1973) has also provided a survey of the use of *Drosophila* in the investigation of developmental problems. Probably no other eukaryotic organism has been studied in equal detail from both genetic and developmental perspectives.

Genetic Organization and Fine Structure

A fairly detailed description of how genetic information is organized in a typical eukaryotic genome is likely to be important for our understanding of how expression of the information is regulated, as well as for complete understanding of processes related to gene transmission, such as replication and recombination. A number of simple and complex gene loci have been studied in considerable detail in *Drosophila*. It is often the case that reasonable models of the organization and functional interactions within a given region can be developed on the basis of genetic information (e.g., LEWIS, 1964). However, critical tests of such models are likely to depend on biochemical information. Clearly, such information will be easier to obtain if the biochemical nature of the gene product is known at the outset, as with defined gene-enzyme systems. The detailed studies on the organization of the *ry* and *ma-l* loci, and on the nature of recombination and gene conversion events at these loci (pp. 53 and 57), provide excellent examples of what is possible.

Representative of the questions needing answers are those raised by the recent indications (e.g., JUDD *et al.*, 1972) that each chromomere in a polytene chromosome contains a single functional unit, even though there is sufficient DNA for many structural genes. These observations invite the speculation that each chromomere includes regulatory components as well as structural information. Indeed, several models of gene regulation in eukaryotes call for this kind of organization (e.g., BRITTEN and DAVIDSON, 1969; GEORGIEV, 1972). Direct molecular characterization of the genome (see LAIRD, 1973) can contribute to our understanding of organization at this level, but detailed genetic studies on well defined loci are also likely to be important.

Population Genetics and Evolution

Allelic variations in enzymes, particularly electrophoretic variants, have provided population geneticists with a much needed tool for measuring the genetic variability in populations (HUBBY and LEWONTIN, 1966; LEWONTIN and HUBBY, 1966; JOHNSON *et al.*, 1966). An extensive literature has developed in this area over the last decade, and detailed treatment of the problems and approaches in this field is beyond our scope. A recent review may be found in LEWONTIN (1973). Comparisons between enzymes in different species have also been used as a taxonomic tool. A few examples of this application are referred to in subsequent chapters (pp. 117 and 128).

Biochemistry, Physiology, and Metabolic Pathways

Mutants unable to carry out a given overall biochemical process or pathway have been key tools in dessecting complex or multi-step processes in many organisms. The analysis of the ommochrome pathway given in Chapter 2 is the best example of this approach in *Drosophila*. The work on pterin pigments, also discussed in Chapter 2, indicates some of the problems and pitfalls. When and if extensive series of auxotrophic mutants become available, they should contribute to our understanding of other pathways. *Drosophila* has not been particularly popular for work on insect physiology, probably because of the small size, but the genetic tools available might compensate in many cases.

Methods

Genetic Methods

Classes of Useful Mutants

It is essentially trivial to point out that no genetics is done without mutants, or perhaps more properly, without the availability of alternative alleles at the locus to be studied. Hence, finding appropriate genetic variants is a necessary first step in any genetic characterization of a gene-enzyme system. A number of classes of mutations are potentially useful under various circumstances.

Genes With a Visible Phenotypic Effect Caused by a Known Enzymatic Defect. Probably a large proportion of all the known visible mutations in *Drosophila* cause a deficiency in one or more enzymes. However, in only a handful of cases is the specific enzymatic defect known (see Chapter 2). Getting from a visible phenotype to a biochemical defect generally has been extremely difficult, and there is no assurance that even a major effort in this direction will be successful. However, in those cases where a relationship has been established, the visible phenotype can greatly facilitate the study of the gene-enzyme system. The very detailed studies on the *rosy (ry)* cistron, a structural gene for xanthine dehydrogenase, were made possible in part by the ability to recognize rare mutational and recombinational events through their visible phenotypic expression (see Chapter 3). Genes which control some of the steps in ommochrome synthesis, mutations of which lack the brown eye pigments, have also been studied in considerable detail (p. 34).

Enzyme "Null" Mutants without Associated Visible Phenotypes. A number of mutants which lack detectable levels of specific enzymes, but which appear to be perfectly normal as far as visible phenotype is concerned, are now known. These are typically more difficult to work with than the preceding class, because flies must be ground up and assayed to determine their phenotype and genotype. In principle, it should be possible to escape this difficulty in at least some cases by devising chemical selection procedures to dinstinguish between mutant and normal flies. This approach is discussed in more detail below (p. 19).

Electrophoretic Variants. The application of histochemical staining methods to enzymes separated by electrophoresis in gels or other supporting media (see HUNTER and MARKERT, 1957; SHAW, 1965), opened up a whole new approach to the problem of obtaining a genetic handle for use in studying enzymes. Ever since WRIGHT (1961) discovered a genetically controlled eletrophoretic variant of an esterase in *Drosophila*, a steady stream of electrophoretic variants of a wide range of enzymes has been reported. The two great advantages of electrophoretic variants are that they occur relatively commonly among wild and laboratory populations, and that the flies carrying variant enzyme forms are assumed to be essentially normal physiologically (but see p. 119). This means that if an enzyme has been selected for study because of interesting biochemical, physiological or developmental properties, there is a good chance that suitable variants can be found to allow genetic characterization, even if null mutants might be lethal.

Mutations with Quantitative or Qualitative Effects on Enzyme Expression. We have already mentioned the potential interest of mutations which affect the expression of a given enzyme rather than its structure (p. 4). Such mutations may serve to identify regulatory loci and provide tools for investigating the nature of the regulatory mechanisms. Only a few such mutations have been described, but systematic searches have begun only recently (p. 10).

Temperature-Sensitive Mutations. Conditionally lethal mutations, of which temperature-sensitive (ts) lethals are the best known class, offer an alternative way of studying genes with indispensable functions. Since the work on ts lethal mutations by SUZUKI *et al.* (1967), there has been considerable interest in such mutations as tools in the study of a number of problems. Temperature shift experiments may permit one to define periods when the gene product is required. An additional advantage follows from the fact that ts lethals may be carried in homozygous condition and, following a temperature shift, all individuals in the population will express the mutant phenotype. This should permit biochemical analyses not possible with ordinary lethals, which must be carried in balanced stocks in which only 25% of the population is homozygous for the mutation of interest (see WRIGHT, 1968, 1970 for further discussion of this point).

In principle, ts mutations affecting specific enzymes could be obtained. However, the usual screening methods identify ts mutations solely on the basis of the lethal phenotype. The same difficulties encountered in tracing mutations with visible phenotypes to primary biochemical defects apply to ts lethals, and none have been traced so far. It may be possible to obtain ts alleles at previously defined enzyme loci at which there exist non-ts null alleles, and for which efficient screening systems exist (an easily scored visible phenotype or a chemical screening system). Such mutations might prove very useful both in understanding the nature of ts alleles and in analyzing the developmental expression of the particular enzyme locus. For a more complete discussion of the isolation, properties and uses of ts mutations, see the review by SUZUKI (1970).

Obtaining Useful Mutants

In designing a procedure to screen for mutants useful for studying gene-enzyme systems, two things must be considered. How will the necessary genetic variants be generated, and how will they be identified and picked out of the population? We will consider sources of variability first, assuming for the moment that detection methods applicable to single flies are available. We will then take up the methods of detection.

Naturally Occurring Variants

A considerable body of literature exists concerning the nature, extent and significance of genetic variability of enzymes in *Drosophila* populations. Of particular interest to us is the conclusion, now well documented, that enzyme polymorphisms are extremely common and rather easy to find in most species of *Drosophila* (HUBBY and LEWONTIN, 1966; LEWONTIN and HUBBY, 1966; JOHNSON *et al.*, 1966; O'BRIEN and MACINTYRE, 1969; AYALA *et al.*, 1972). Typical results indicate that electrophoretic polymorphism exists at more than half of all enzyme loci, and that the frequency of the less common alleles is high enough to allow detection in surveys of reasonable size. While the probability of finding variability at any given enzyme locus is rather good, there is some evidence that variants of enzymes involved in important metabolic pathways may be less common (GILLESPIE and KOJIMA, 1968; KOJIMA *et al.*, 1970; AYALA and POWELL, 1972).

Naturally occurring null alleles not associated with visible phenotypes have been discovered at several enzyme loci, including those for aldehyde oxidase (p. 56), pyrodoxal oxidase (p. 57), leucine amino peptidase-A (p. 115), esterase-C (p. 119), and larval alkaline phosphatase (p. 123). In most cases they were discovered in conjunction with electrophoretic surveys when a band of enzyme activity was found to be missing rather than just altered. Null mutants are certainly much rarer than electrophoretic variants in natural populations, and surveys whose sole purpose is to discover null mutants are almost certain to be laborious and rather likely to fail.

Genetic variants affecting the quantitative or qualitative expression of enzyme loci have not been extensively searched for, but enough evidence is available to suggest that considerable variability of this type exists. DICKINSON (1971, 1972, 1974; see also p. 66) has found considerable variation between stocks in the developmental timing and tissue distribution of aldehyde oxidase. KELLER and GLASSMAN (1964b,c; see also p. 61) found a wide range of quantitative differences in xanthine dehydrogenase during an extensive stock survey. Most of the variation was attributable to genetic differences (KELLER, 1964), and at least one well defined locus (*low xanthine dehydrogenase;* see p. 61) was identified in this way. KING (1969) found variations in the developmental expression of xanthine dehydrogenase among inbred strains of *Drosophila*. Strain specific quantitative or qualitative differences have been reported for protease (p. 115), sucrase and trehalase (p. 132). Slightly removed from the enzyme level, but suggesting similar variability, a number of workers have noted strain specific variations in the pterin content of specific tissues and organs (p. 48).

Obtaining Homozygous Stocks

Once useful, naturally occurring genetic variants are identified in a population, further genetic work generally requires establishment of strains monomorphic for each allele. In many surveys in which a series of laboratory populations is examined such strains will be detected in the survey. The same is true when a series of stocks established from single wild-caught females is used (see HUBBY and LEWONTIN, 1966). When a desired allele has been detected only in polymorphic stocks, monomorphic stocks may be established by inbreeding (pair matings), or by use of one of the standard schemes for rendering individual chromosomes homozygous. A generalized representation of this kind of system is shown in Fig. 1. This approach has the disadvantage that a minimum of three generations is required, and many separate sublines must be handled. Furthermore, without prior knowledge about which chromosome carries the locus in question (knowledge that is generally only obtained after monomorphic stocks are established), one may have to test a series of stocks rendered homozygous for each chromosome, or use a more complex system to render all chromosomes homozygous at once. For these reasons, simple inbreeding is usually the simplest approach. A series of pair matings is set up, and subsequently either the parental pair or a sample of the progeny are recovered and tested to see which allele(s) are present. Fewer individual flies will need to be tested if the parents are tested rather than the progeny. Even when an allele has been found in an apparently homozygous stock, it may be advisable to establish sublines from pair matings and test the parents to rule out the presence of rare alleles.

The number of pair matings that must be set up and tested to give reasonable assurance of recovering a desired allele in a homozygous stock is easily calculated once the relative frequency (r) of that allele is known. The probability (p) that a pair mating will have both parents homozygous for the desired allele is r^4. The probability of failing to find a suitable pair among n trials is $(1-p)^n$. It can easily be calculated that even for alleles as commom as 20–30% of the gene pool, the number of matings that must be set up is rather large. For allele frequencies much smaller than this, it may be necessary to go through a stepwise process of selective breeding, using the progeny of the mating with the highest frequency of the desired allele as parents for the next generation of pair matings. The process may be repeated as often as necessary. Electrophoretic variants may be rendered homozygous in this way, but in addition, it seems possible to recover various quantitative and qualitative variants. KELLER and GLASSMAN (1964c) were able to recover the mutant *lxd* in this way, and CHAUHAN and ROBERTSON (1966; see also p. 48) were able to produce strains with significant specific biochemical differences.

Induced Mutations

Induced mutations have been obtained thus far at only a few defined enzyme loci, and in most cases the locus had been previously identified using natural variants. Examples can be found in the sections on alcohol dehydrogenase (p. 75), α-glycerophosphate dehydrogenase (p. 87), xanthine dehydrogenase (p. 53), and acid phosphatase (p. 129). The existing work is sufficient to demonstrate that this approach is useful for either obtaining new classes of mutants or a more extensive series of alleles at a defined locus. In principle, it should be possible to obtain mutants affecting enzymes for which no natural variants have been found.

In the above cases, the new mutants at enzyme loci were induced either with X-rays or with the chemical mutagen ethyl methanesulfonate (EMS). Presumably, other chemical mutagens could be used. Detection of newly induced recessive mutations normally involves procedures to render the mutagenized chromosomes homozygous (Fig. 1). These rather time-consuming procedures have generally been avoided in the work with enzyme loci. Induced electrophoretic variants are detectable in a fairly direct way (Fig. 2). Provided both parents used in a cross have the same electrophoretic form of an enzyme, an induced variant will be detectable as an altered electrophoretic pattern when single F1 flies are analyzed electrophoretically (p. 17). However, it is clearly necessary to mate each fly individually to preserve the mutant chromosome prior to the electrophoretic analysis. Normally this mating employs partners with markers and balancers chosen to facilitate recovery of the

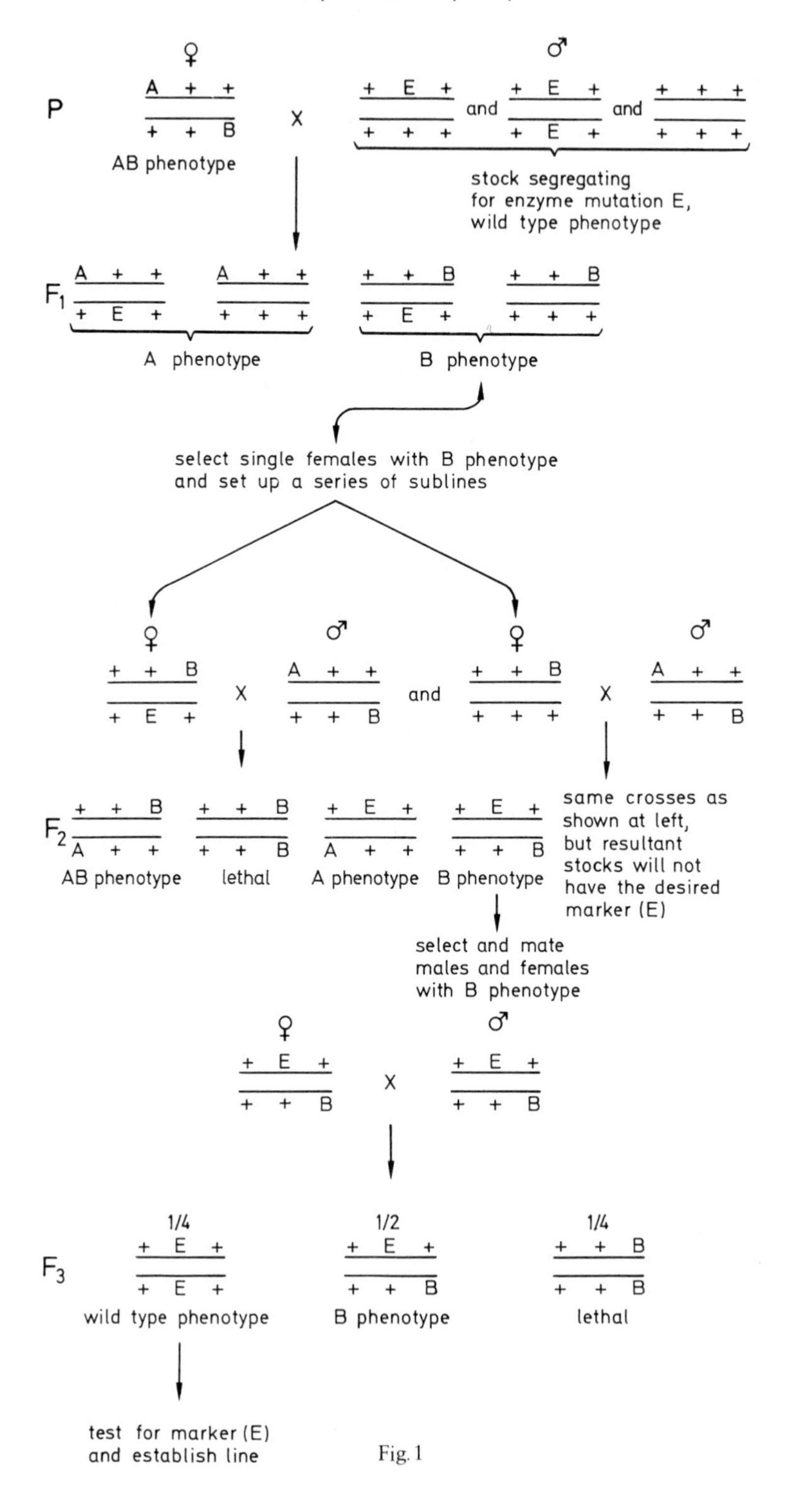

P
A + +
+ + B
AB phenotype
X
+ E +
+ + +
and
+ E +
+ E +
and
+ + +
+ + +
stock segregating for enzyme mutation E, wild type phenotype
F1
A + +
+ E +
A + +
+ + +
A phenotype
+ + B
+ E +
+ + B
+ + +
B phenotype
select single females with B phenotype and set up a series of sublines
+ + B
+ E +
X
A + +
+ + B
and
+ + B
+ + +
X
A + +
+ + B
F2
+ + B
A + +
AB phenotype
+ + B
+ + B
lethal
+ E +
A + +
A phenotype
+ E +
+ + B
B phenotype
same crosses as shown at left, but resultant stocks will not have the desired marker (E)
select and mate males and females with B phenotype
+ E +
+ + B
X
+ E +
+ + B
F3
1/4
+ E +
+ E +
wild type phenotype
1/2
+ E +
+ + B
B phenotype
1/4
+ + B
+ + B
lethal
test for marker (E) and establish line

Fig. 1

mutagenized chromosome, should it prove to carry the desired mutation. It also appears that it may be necessary to screen up to several thousand individuals, each mated separately, to obtain a new variant at a specific locus (see pp. 75 and 129). Thus, considerable effort is involved, and naturally occurring variants are likely to be easier to find in most cases.

Given at least two distinguishable electrophoretic alleles, it is possible to screen for null alleles (Fig. 3). The procedure, which again invoies electrophoretic screening of previously mated individuals, is based on using male and female parents with different electrophoretic alleles, and looking for progeny in which only one parental form is detectable in F1 flies (see p. 129). If at least one null allele is known or has been obtained as above, additional null alleles at the same locus can be detected without electrophoresis, if a convenient spot assay for the enzyme is available (BELL *et al.*, 1972; see also p. 129). It is still necessary to mate each individual before doing the analysis (Fig. 4). As an alternative, once a locus is defined by one kind of mutation, and its map position is known, it may be possible to obtain a deletion which includes the enzyme locus. In any subsequent search for new mutations, mutagenized chromosomes may be made heterozygous with the chromosome carrying the deletion in a single cross, and any new mutation within the region exposed by the deletion will be expressed. This strategy is illustrated in the section on alcohol dehydrogenase (p. 75).

Enzyme mutants with an associated visible phenotype are clearly easier to work with in this respect. Newly induced mutations may be identified visually and mated after identification. It is therefore possible to screen much larger numbers of mutagenized chromosomes. Appropriate chemical screening system can facilitate testing of even larger numbers or progeny. We shall discuss this further below (p. 19).

One final difficulty with induced mutations should be mentioned. Mutagen doses that produce mutations in the target locus at a useful frequency (say one in a few thousand treated chromosomes) are almost certain to induce numerous other mutations, including recessive lethals, at other sites on the same chromosome (BELL *et al.*, 1972). If the planned use of the induced mutations involves rendering them homozygous, it will be necessary first to cross out all linked recessive lethals. This is likely to be a tedious and time-consuming process (BELL *et al.*, 1972).

Fig. 1. Schematic representation of a system for rendering a chromosome homozygous. In this system, A and B represent dominant mutations controlling morphological characters and also behaving as recessive lethals. B is further assumed to be on a chromosome carrying multiple rearrangements that suppress recombination. A/B therefore constitutes a balanced lethal system. E represents a mutation affecting some enzyme, but with no visible morphological effect. Note that from the second generation on, each manipulation must be done on multiple individual sublines, each derived from a single female. The proportion of "successful" sublines (flies homozygous for E recovered in F_3) is a function of the frequency of the E allele in the starting population. Furthermore, it is probable that many of the chromosomes in the starting population will carry recessive lethals, and no phenotypically wild-type flies will be recovered at F_3. Hence, the number of sublines which must be started to obtain some E/E lines may be quite large. Finally, we have assumed here that the enzyme marker is on the same chromosome as A and B. If this is not the case, the phenotypically wild-type flies at F_3 will still be segregating for the enzyme marker

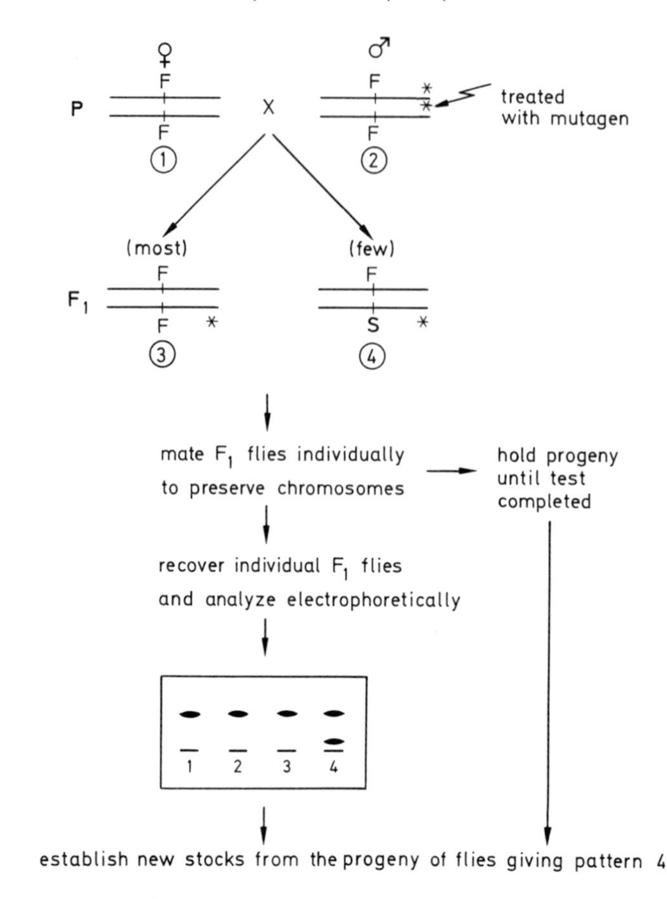

Fig. 2. Induction of new electrophoretic alleles. F and S represent respectively a naturally occurring enzyme structural gene and an induced electrophoretic variant. The asterisk (*) indicates chromosomes that have been exposed to the mutagen. The rectangle near the bottom represents an electrophoretic gel. The origin is indicated by four lines representing sample slots. The numbers under the sample slots correspond to the numbers under the diagrams of genotypes. The occurrence of a new band in sample 4 indicates the production of a mutant allele

Methods of Detection

Identification of mutant individuals in a population is central to both the discovery of useful mutations which we have been discussing, and to their subsequent use in a variety of contexts to which we shall turn shortly. A suitable method of identification should be sufficiently easy and rapid to permit large numbers of analyses in a reasonable time, and sufficiently sensitive to be applicable to single flies. Visible mutations obviously present no difficulties in this respect, so the following discussion deals with those mutations defined only at the enzymatic level.

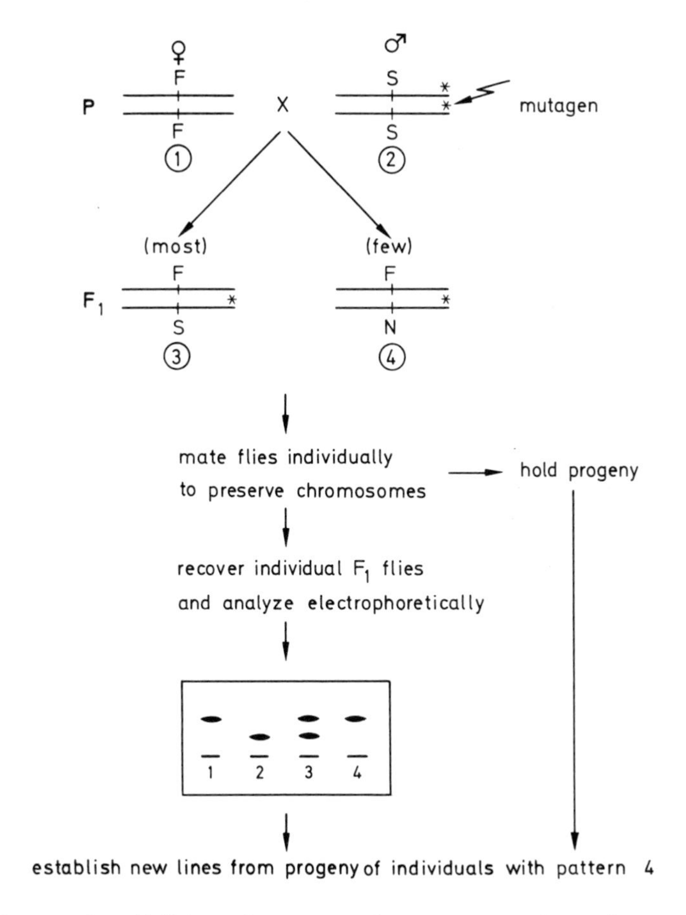

Fig. 3. Induction of a null allele. In this case, F and S represent two electrophoretic alleles, either naturally occurring or induced. N represents an induced null allele at the same locus. In the F 1 generation, most individuals will express both parental forms as in sample 3. Absence of the parental form suggests presence of a null allele, although a mutation to an allele yielding an enzyme with electrophoretic mobility like the maternal form is possible. Subsequent tests on the line derived from the progeny easily distinguish between these possibilities

Electrophoretic Methods

It is beyond the scope of this work to provide detailed accounts of the wide variety of electrophoretic separation and staining methods now available. Methods commonly used with specific enzymes are referenced in the appropriate chapters. Several general sources may also be mentioned. The volume edited by WHIPPLE (1964) contains accounts of methods and applications for various forms of gel electrophoresis. SHAW and PRASAD (1970) have recently compiled an extensive series of electrophoretic procedures and staining recipes useful in starch gels. The

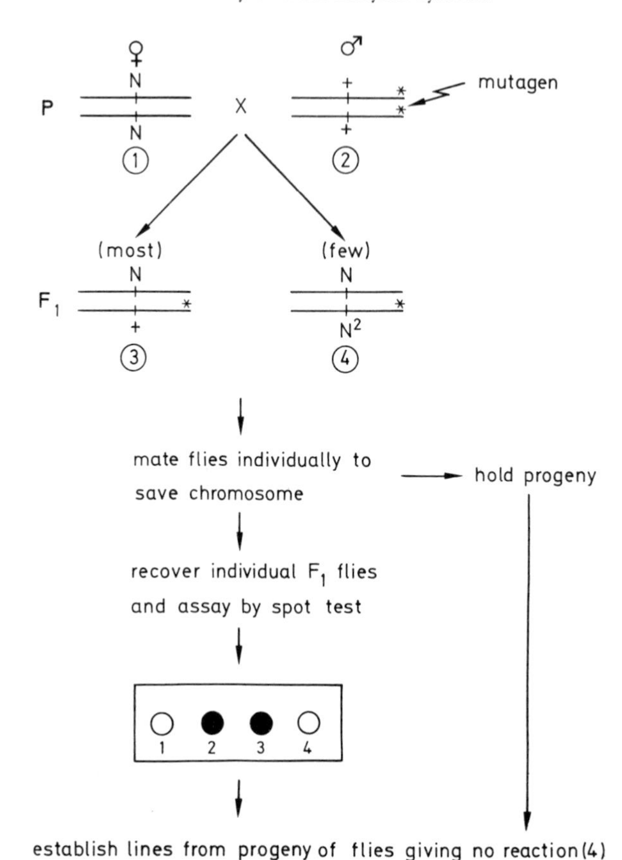

Fig. 4. Induction of additional null alleles. N represents a previously identified null allele and N^2 a newly induced null. In the rectangle near the bottom, the circles represent spot tests for enzyme activity, with numbers corresponding to the numbers below each genotype diagrammed above. An empty circle represents a negative test (no activity), and a solid circle represents a positive test

stains should be applicable to other media as well. JOHNSON *et al.* (1966) list a number of starch gel methods specifically applicable to *Drosophila.* Likewise, KNOWLES and FRISTROM (1967) and HUBBY and LEWONTIN (1966) list acrylamide gel methods for several enzyme systems in *Dorsophila.* URSPRUNG and LEONE (1965) have described an agar gel method for *Drosophila* alcohol dehydrogenase that has subsequently been extended, with appropriate staining methods, to a number of other enzymes.

Most electrophoretic methods have the advantage of requiring small samples, usually only a few microliters, and most of the methods referred to above are applicable to extracts of single flies. The methods now available permit screening of hundreds or even thousands of individual flies in reasonable times. It should be noted that in some cases it is possible to stain simultaneously or sequentially for more than one enzyme, or to stain slices of the same gel for different enzymes and thus get more information from the invested time.

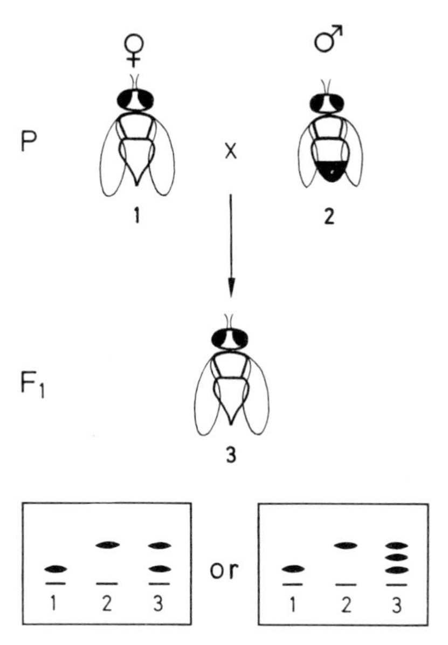

Fig. 5. Two patterns of codominant expression. In crosses between parents with electrophoretically different forms of an enzyme, the F 1 flies may give a pattern which is a simple sum of the parental patterns (left hand gel), or may have one or more additional bands (right hand gel)

One of the great advantages of the electrophoretic method is that electrophoretic alleles appear to be almost universally co-dominant in expression. Heterozygotes usually have a pattern that represents a simple sum of the parental patterns or a pattern that includes some new "hybrid" forms in addition to the parental bands (Fig. 5). Hybrid bands are generally interpreted as indicating that the enzyme in question contains multiple subunits coded by the same locus. Biochemical confirmation of this model is available for some enzymes, including 6-phosphogluconate dehydrogenase (p. 85), and acid phosphatase-I (p. 127). This co-dominant expression means that rare or newly induced mutations may be detected without rendering them homozygous (p. 11).

A number of situations other than existence of multiple alleles at a structural locus can lead to multiple electrophoretically separable bands of enzyme activity. It might be appropriate to digress briefly and discuss some of these. 1. Products of distinct, non-allelic genes may have similar or overlapping substrate specificities and hence stain with the same reaction mixture. This is probably the explanation for the complex, multiple-banding patterns commonly found with peptidases (p. 114), esterases (p. 116), and phosphatases (pp. 122 and 127). In some cases, the genes may have arisen by duplication followed by diversification. Multiple amylases in *D. melanogaster* (p. 110) and aldehyde oxidases in *D. nhdei* (p. 56) may represent relatively recent gene duplications. 2. A series of isozymes may be generated when active enzyme molecules are made up of variable combinations of subunits coded by two or more different loci, as in the classic case of vertebrate lactate dehydrogenase (MARKERT, 1963). No well established example is known in *Drosophila*. 3. Products of the same gene may be secondarily modified to alter the electrophoretic mobility. Apparent examples of this case in *Drosophila* include alcohol dehydrogenase (p. 72), xanthine dehydrogenase (p. 51), and an alkaline phosphatase (p. 126). 4. Allelic variants at an enzyme locus may be responsible, as discussed above.

Genetic tests often facilitate distinguishing between these alternatives. To demonstrate case 4, examination of single flies or inbred lines should reveal two or more genetically segregating patterns. In cases 2 and 3, if a segregating allelic difference affecting one band can be found, the other related bands will be shifted coordinately. Examples like this in *Drosophila* include alcohol dehydrogenase (p. 72), hexokinase (p. 133), aldehyde oxidase (p. 52), α-glycerophosphate dehydrogenase (p. 87), and an alkaline phosphatase (p. 126). Rigorous distinctions between cases 2 and 3 require some biochemical analysis. The discussions of alcohol dehydrogenase (p. 72), octanol dehydrogenase (p. 78), and α-glycerophosphate dehydrogenase (p. 87) provide examples of approaches to this problem. Evidence for case 1 may often be obtained by experiments on substrate specificity and sensitivity to inhibitors. Examples of this approach include studies on hexokinases (p. 132), esterases (p. 117), and the alcohol dehydrogenase—octanol dehydrogenase system (p. 71). We should also point out that enzymes may differ in other properties, even though they are electrophoretically indistinguishable. Many amino acid substitutions are possible that do not alter the net charge. Cases are known in *Drosophila* of allelic variants with different heat stabilities but identical electrophoretic mobility (esterase-6, p. 119), and of possible non-allelic genes which produce enzymes of identical electrophoretic mobility (hexokinase, p. 132).

Quantitative Methods

Quantitative assay methods applicable to many of the enzymes discussed in this book are referenced in the respective chapters. Methods useful for enzymes from *Drosophila* have frequently been adapted from methods used with enzymes from other sources. Any standard reference work on enzymes can serve as a source for such methods. As indicated above, for genetic purposes it is desirable that assay methods be simple, fast and sensitive enough to be applied to single flies. Many of the standard quantitative assays do not satisfy these requirements, particularly as to simplicity and speed. Most are adequate for comparing various strains for quantitative differences, as in the stock surveys comparing activities of xanthine dehydrogenase (p. 61), pyridoxal oxidase (p. 57), protease (p. 115), sucrase and trehalase (p. 132), and cytochrome oxidase (p. 142).

For the rapid analysis of large numbers of individual flies, it may be possible to modify either a spectrophotometric assay or an electrophoretic stain for use as a spot assay. Single flies may be ground in a small volume of buffer, and the reaction mixture added. The change in color is estimated visually to give a qualitative or semi-quantitative measure of enzyme activity. Null mutations, if available, provide useful controls for demonstrating the specificity of such methods. Spot assays have been used in genetic studies on aldehyde oxidase (DICKINSON, 1970) and acid phosphatase (BELL *et al.*, 1972; see p. 129).

Reference has already been made to the fact that null mutants can be recognized by the absence of an enzyme band in an electrophoretic analysis. It might also be mentioned that in some cases where non-specific reactions due to other enzymes and endogenous substrates make a simple spot assay impractical (dehydrogenases may be prone to this problem), an electrophoretic separation can remove the background.

Chemical Selection Systems

The ability to select positively between mutant and normal individuals in a large population on the basis of differential survival in various media has been one of the most powerful tools in microbial genetics. Comparable methods are now becoming possible with *Drosophila,* and promise to make feasible mutant screens and genetic experiments on a scale and with a resolution not otherwise possible.

At least three general strategies are possible: 1. One can expose the flies (or larvae) to a chemical that is toxic unless the substance can be metabolically altered in some way. Mutants lacking the necessary enzyme(s) will be killed. 2. One can raise flies on a medium lacking an essential metabolite but containing some appropriate precursor. Individuals lacking the enzymatic machinery to synthesize the missing metabolite (auxotrophs) will be selected against. 3. One can expose flies to a chemical that is relatively non-toxic, but that can be enzymatically converted to a toxic substance. This approach, unlike the others, selects in favor of mutant flies. Thus, the last approach holds the greatest promise for screening for new mutants. Use of the first two approaches for this purpose involves some sort of replica-culture method to pick up cultures that survive under standard conditions, but not when the selective conditions are applied. All three methods are potentially useful for genetic analysis once mutants are available. An ideal system would provide for selection in either direction.

Thus far, good chemical selection systems have been devised and applied to two gene-enzyme systems in *Drosophila.* Flies lacking xanthine dehydrogenase are killed when raised on a medium containing purine (p. 56). This procedure has contributed substantially to the fine structure analysis of the *ry* and *ma-l* loci, and to the studies on the nature of recombinational events in those loci (see pp. 53 and 57). This method has not been useful in obtaining new mutants, since it selects against flies lacking the enzyme.

Selective systems working in both directions are available for alcohol dehydrogenase (p. 75). Ethanol at high concentration kills null mutants, but now wild-type flies. Certain unsaturated secondary alcohols select against wild-type flies. This method has been used to obtain new null mutants (SOFER and HATKOFF, 1972; see also p. 76). It should be noted that the development of the systems devised so far depended on existing null mutants for use in testing and calibration. It may be difficult to work out comparable selection methods for use on gene-enzyme systems that are not already fairly well characterized. There also has been some work done on selecting *Drosophila* auxotrophs (VYSE and NASH, 1969; NORBY, 1970; VYSE and SANG, 1971; NASH and FALK, 1973; FALK and NASH, 1973). While this method holds considerable promise, at the time of writing no mutants defined at the enzymatic level have been reported in the literature.

The main use of the systems described above has been more detailed analysis of the known structural loci. There is some hope that these methods can be used also to study regulatory loci, perhaps even those with qualitative or quantitative effects short of total loss of the enzyme.

Genetic Analysis of Mutations Affecting Enzymes

The wealth of genetic studies on *Drosophila* and the sophistication of the genetic manipulations possible with this organism contribute greatly to the potential usefulness of the defined gene-enzyme systems considered in this book. It is unfortunate that there exists no unified account of *Drosophila* genetics, or of methods and procedures useful in the study of *Drosophila* genetics, but it is far beyond the scope of this book (and the expertise of its authors) to provide such an account. Therefore, in the following pages we shall consider only a few problems and methods that are specifically relevant to work with genes defined only at the enzymatic level. Those mutations with known enzymatic defects and visible phenotypic effects may be handled in conventional ways.

General Considerations

Mutations whose only apparent effect is at the enzyme level present at least one major difficulty, but also have an advantage that partially offsets the difficulty. The difficulty is obiously their "invisibility", i.e., flies must be ground up and assayed to determine phenotypes. This has two important implications. Firstly, more work is involved and hence fewer flies may be examined. Secondly, the flies are destroyed in the process and may not be used for further crosses after the phenotype is determined. If further crosses are required by the experimental design, they must be set up blind in advance. If the desired phenotypic class is present as only a small fraction of the total population, a great many unwanted crosses must be set up (and the flies involved tested), to be sure of getting at least some of the desired crosses.

The one important advantage of many enzyme mutations (probably most structural gene mutations) is that they are co-dominant in expression. By this we mean that both structural genes in a diploid cell are expressed independently. Therefore, heterozygotes in crosses involving allelic enzyme mutations have a phenotype in some way intermediate to the parental phenotypes, and distinguishable from both parents. For example, we have mentioned that heterozygotes produced in crosses between parents with different electrophoretic forms of an enzyme generally have an electrophoretic pattern that includes both parental forms and, sometimes, additional hybrid bands (p. 17). There is also considerable evidence that the level of enzyme activity in heterozygotes involving null alleles is normally about half of the wild type level. This is best documented with respect to alcohol dehydrogenase (p. 74), xanthine dehydrogenase (p. 53), tryptophan oxygenase (p. 39), kynurenine hydroxylase (p. 42), and α-glycerophosphate dehydrogenase (p. 87), but is probably a rather general phenomenon. This co-dominant behavior means that the genotype may be inferred directly from the phenotype. This in turn means that test crosses or other extra steps commonly used to determine genotype when recessive mutations are being used are unnecessary. Furthermore, in most kinds of genetic experiments, heterozygotes will be more common than homozygotes, so the number of operations "wasted" because of working blind will be reduced, often quite significantly.

Identification of Structural Loci

For some purposes it is necessary to distinguish between structural genes which code for enzyme primary structure, and other loci that influence expression of enzyme activity. The best criterion is that at least some mutations in a structural locus will produce altered protein products. Alleles that control electrophoretic differences are commonly assumed to define structural loci, and this is probably usually true. However, enough evidence exists for outside factors that modify electrophoretic mobility (see pp. 51, 72, and 126) to suggest caution. Demonstrations that null mutants and electrophoretic variants are controlled by the same locus strengthens the case. The test for allelism of null and electrophoretic variants involves crossing null homozygotes to homozygotes of each electrophoretic type. If the mutations are allelic, only the electrophoretic variant present in the positive parent will appear in the F1 flies from each cross (Fig. 6). Demonstrations of co-dominant expression of electrophoretic variants and/or null mutants also suggest that the locus in question is a structural locus. The demonstration that the *ry* locus is a structural gene for xanthine dehydrogenase (p. 53) provides a good illustration of the combination of criteria that can be applied.

Assignment to a Linkage Group

The co-dominant expression of enzyme mutations typically makes assignment to a linkage group rather easy. Sex linkage is indicated when only females display the co-dominant pattern, and males express only the maternal parental pattern in reciprocal crosses (Fig. 7). Glucose-6-phosphate dehydrogenase (p. 82) and 6-phosphogluconate dehydrogenase (p. 84) provide clear examples of this kind of segregation. Autosomal linkage may be followed with either dominant or recessive visible markers. In the former case, a cross is made between a stock carrying a dominant autosomal marker (usually heterozygous and often in a balanced system that prevents recombination), and a stock homozygous for an allele controlling a different enzyme variant. F1 progeny that carry the dominant marker are backcrossed to the parent with the enzyme variant. F2 flies with and without the dominant marker are analyzed separately. If the enzyme marker is on the same chromosome as the dominant marker, all flies with the marker phenotype will be heterozygous for the enzyme variant, and all flies not expressing the dominant marker will be homozygous at the enzyme locus. If the enzyme locus is unlinked to the marker, homozygotes and heterozygotes will be found at random in both phenotypic classes. Taking advantage of the absence of recombination in *Drosophila* males, the dominant marker need not be balanced to prevent recombination if F1 males are used.

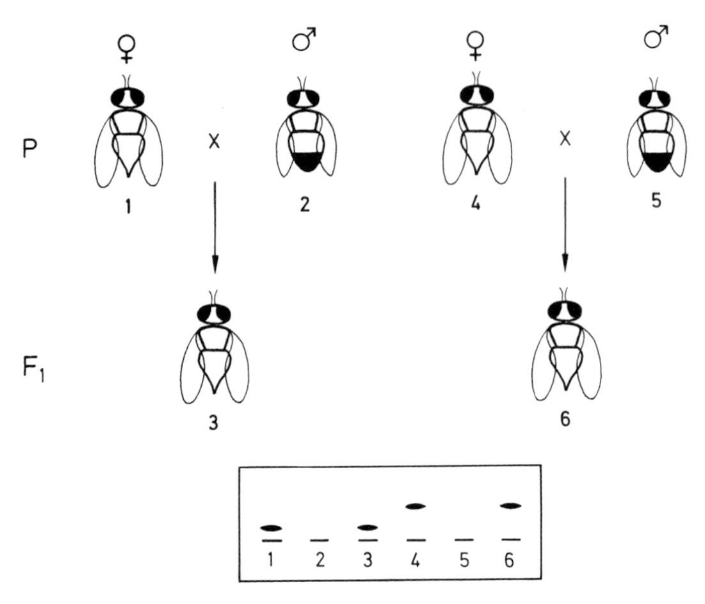

Fig. 6. Allelism of a null mutant to electrophoretic variants. At the top are diagrammed crosses of flies carrying a null mutation to flies with two different electrophoretic forms of the enzyme. At the bottom are shown the electrophoretic patterns. The presence in the F 1 of only the form present in the positive parent in each cross indicates that the null flies are mutant at the same locus as the one controlling the electrophoretic difference

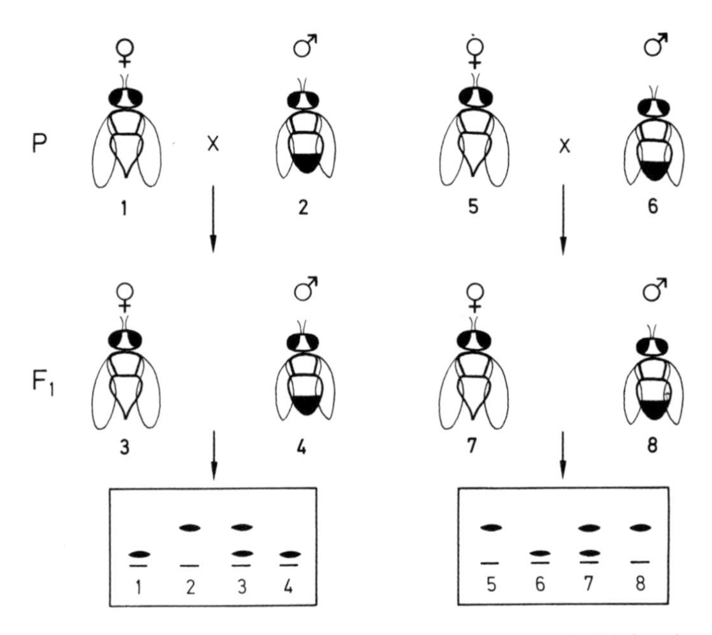

Fig. 7. Segregation of a sex-linked enzyme mutant. In each cross, the F 1 females (3 and 7) have the codominant pattern, while the F 1 males (4 and 8) have only the enzyme form present in their maternal parent

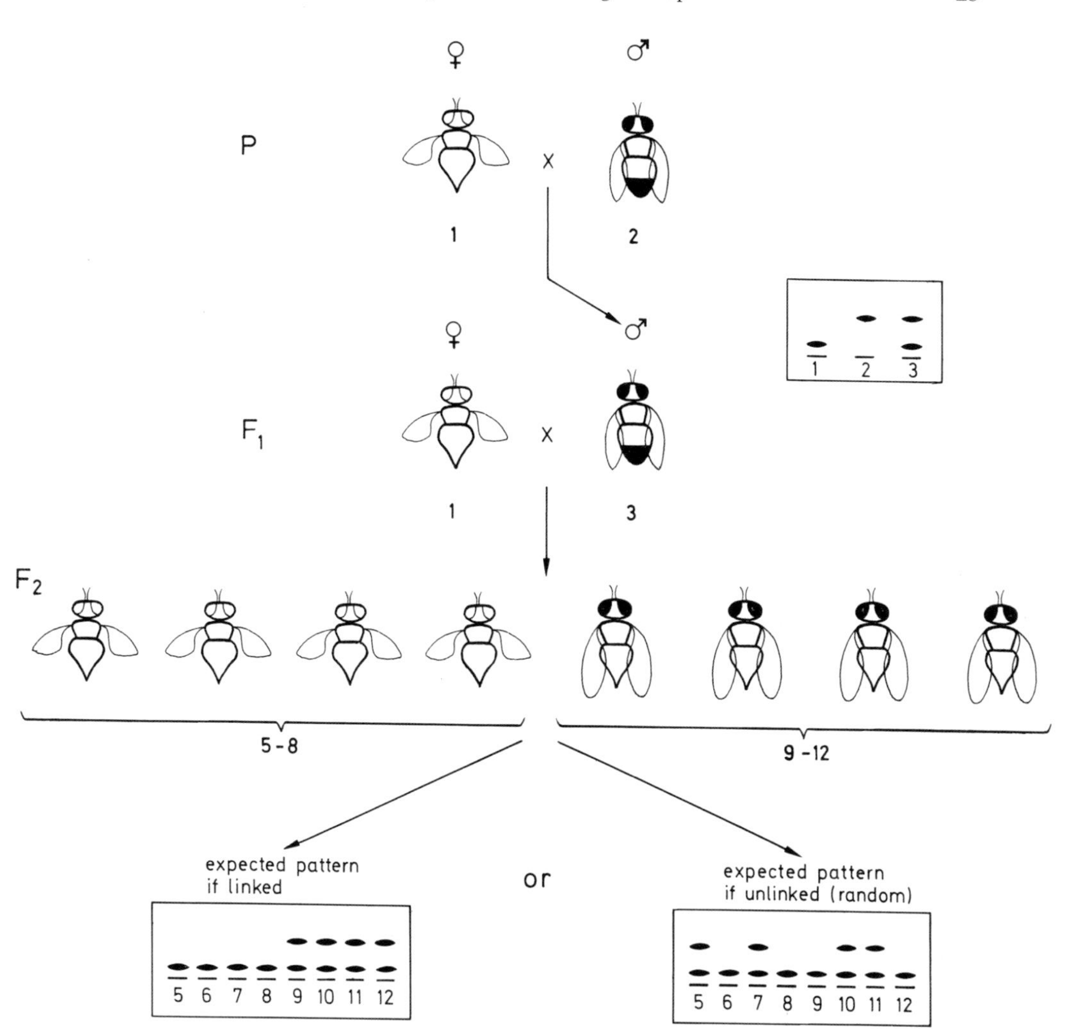

Fig. 8. Autosomal linkage of an enzyme variant using recessive markers. The female parent at the top is shown as carrying an eye color mutation and a wing mutation, both recessive and on the same autosome. The electrophoretic patterns for the two parents and the F1 heterozygotes are shown at the right. F2 flies are sorted by visible phenotype and analyzed electrophoretically. The uniform pattern at the lower left indicates linkage of the enzyme locus to the visible markers. The random pattern at the right indicates no linkage

Linkage to recessive markers is just about as easily established (Fig. 8). Flies homozygous for one or more recessive markers and a specific allele at an enzyme locus are crossed to flies homozygous for a different enzyme allele. F1 males (again to avoid recombination) are backcrossed to the marker stock. Phenotypically mutant and wild-type progeny are analyzed separately. In the case of linkage, all normal flies will be heterozygous for the enzyme variant, and all flies expressing the visible mutant phenotype will be homozygous for the enzyme allele originally present in the marker stock. If the enzyme variant is not detectable in single flies or is recessive (unlikely for a structural gene mutation but possible for a regulatory mutation), the system using dominant markers is preferred. The phenotypically wild-type F 2 flies may be pooled for assay, and are expected to be homozygous for the enzyme mutation if the locus is on same linkage group as the visible marker.

Mapping

Once a linkage group assignment has been made, estimation of a map position is generally based on recombination with recessive markers in a series of crosses similar to the second one described above (Fig. 9). Usually a marker stock with a series of recessive markers on the chromosome in question is used, and F 1 females are used in the backcross to permit recombination. F 2 flies are classified according to the visible markers, and flies from suitable classes are analyzed to determine whether they are homozygous or heterozygous at the enzyme locus. Flies homozygous for any given visible marker can be heterozygous for the enzyme marker only if they carry a chromosome recombinant between that marker locus and the enzyme locus. This clearly provides a way to measure the genetic distance of the enzyme locus from each marker. Consideration of the visible markers in pairs permits assignment of the enzyme locus to a specific chromosomal segment. Selection of flies carrying chromosomes recombinant between a pair of visible alleles that define a chromosome segment provides the traditional three-point cross for refining the map position. Since analysis of the enzyme phenotype is more difficult than classification by visible phenotype, it is generally easier to refine a map position by choosing closer outside markers than by analyzing larger numbers of recombinants. The reader will find examples of this basic mapping strategy throughout the remainder of the book.

The method outlined above hinges on three features of many gene-enzyme systems. 1. The alleles at the enzyme locus are co-dominant; 2. an assay method adequate to classify single flies is available, and 3. the enzyme variant in question is expressed at a developmental stage for which visible markers are also available. If any one of these features is absent it may be necessary to do crosses as above, and classify flies by visible phenotype and then mate them individually to appropriate tester stocks and analyze their progeny to determine their genotype at the enzyme locus. The procedure used by MACINTYRE (1966) to map an electrophoretic variant of larval alkaline phosphatase (see p. 123) illustrates this kind of approach. Similarly, progeny tests were necessary in mapping the *low pyridoxal oxidase* locus, because the assay available was not sensitive enough for single flies (COLLINS and GLASSMAN, 1969). As with assignment to a linkage group, mapping of a recessive gene affecting an enzyme will be facilitated by using dominant markers.

Fine Structure Mapping

Successful resolution of closely linked genes, and analysis of the fine structure within single functional units depends on relatively powerful methods for selecting a small subpopulation that is greatly enriched for the recombinant class(es) being

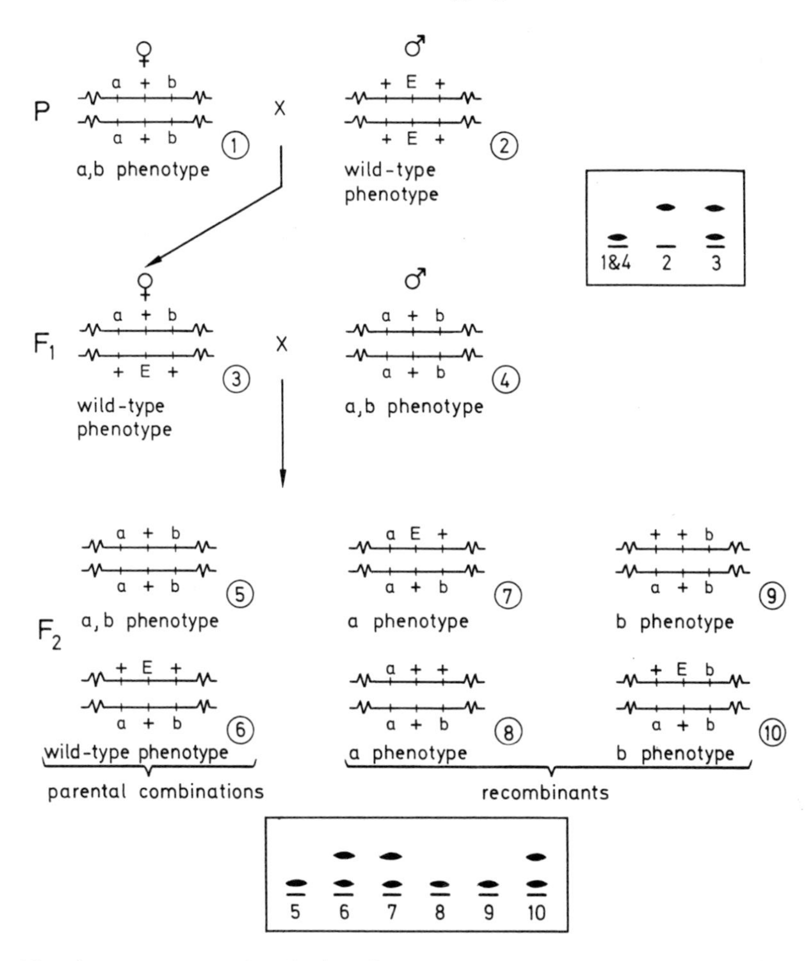

Fig. 9. Mapping an enzyme variant. In this schematic representation, E represents a codominant enzyme variant, and a and b represent recessive markers controlling visible phenotypes. The number next to each genotype corresponds to the numbers by the sample cells in the electrophoretic gels. The electrophoretic patterns for parents and F1 flies are shown at the upper right. Classes of F2 are shown at the bottom. The observation that all F2 flies expressing both visible markers (class 5) have the homozygous enzyme pattern, and all phenotypically wild-type F2 (class 6) have the heterozygous pattern, indicates that the enzyme marker is within (or very close to) the interval bounded by a and b. Among the recombinant classes (7–10), 7 and 9 are recombinant between a and E, and 8 and 10 are recombinant between E and b. The ratio of these two classes establishes the relative distances of a–E and E–b, and hence the position of E if a and b are known. Normally, several markers are used instead of the two shown. They are considered pairwise to determine which segment includes the enzyme marker (chromosomes maintaining parental combinations of the marker pair give patterns like 5 and 6). Recombinants within this interval are then analyzed to determine exact position

sought. The use of closely linked, flanking visible markers is the most direct approach. The separation of two closely linked amylase loci was achieved in this way (p. 110). Even more powerful is the system of flanking lethals devised by CHOVNICK *et al.* (1962; see p. 53). The crosses are done so that, in the absence of recombination, all progeny are homozygous for one of the recessive lethal mutations. Progeny receiving a chromosome recombinant in the desired interval survive. Examination of the relatively small number of survivors gives resolution comparable to that obtained with much larger numbers of progeny from an ordinary cross. Chemical selection methods are more powerful yet for separating mutations within a cistron. The detailed and remarkable analyses of the *ry* and *ma-l* loci (pp. 53 and 57) are by far the best illustrations of the degree of resolution possible. With the development of even better selection systems for mutants at the alcohol dehydrogenase locus (p. 76), similarly detailed studies should be possible there.

Cytogenetic Localization

For some purposes, it is desirable to determine the position of an enzyme locus relative to the characteristic banding pattern evident in the polytene chromosomes present in salivary glands, and a variety of other tissues in *Drosophila*. Cytogenetic loci have been determined for alcohol dehydrogenase (p. 75) and amylase (p. 109) loci, as well as for some of the loci with visible phenotypes (*v*, p. 36 and *ry*, p. 53). Useful deletions or other chromosomal rearrangements are detected by their effects on visible loci linked to the enzyme locus in question, and are then tested to see whether they overlap the enzyme locus as well. A deletion that overlaps (or extends into) an enzyme locus behaves like a null allele in tests with other alleles at the locus. An example is described on p. 75. Recently, STEWART and MERRIAM (1974) have shown that segmental aneuploids can be used to obtain cytogenetic localizations. Flies which contain chromosomal rearrangements that leave them haploid over a small segment of the genome produce reduced levels of any enzyme whose structural gene is located within the haploid segment.

Biochemical Methods

Only in relatively recent years have serious enzymological investigations been pursued using *Drosophila* as an experimental system. This is true because advances in the field of enzymology have permitted more widespread use of the techniques involved, but more importantly because of the realization that detailed characterization of enzymes is an important way in which basic information about genetic structure and function can be obtained.

Purification

To date only six enzymes from *Drosophila* have been purified to the point where there is at least some evidence that homogenous preparations have been obtained (Table 1). There are others which have been purified to the extent indicated in Table 2. In general the procedures applied to *Drosophila* are similar to those applied to enzyme studies in other systems, and generally these have been

Table 1. Enzymes purified from drosophila

Alcohol dehydrogenase	SOFER and URSPRUNG (1968), JACOBSON *et al.* (1970)
Aldehyde oxidase	DICKINSON (1971)
Aldolase	BRENNER-HOLZACH and LEUTHARDT (1968)
β-L-Hydroxyacid dehydrogenase	BORACK and SOFER (1971)
m Malate dehydrogenase	MCREYNOLDS and KITTO (1970)
Trehalase	HUBER and LEFEBVRE (1971)
Xanthine dehydrogenase	SEYBOLD (1973)

Table 2. Partially purified enzymes from drosophila

Enzyme	Fold	Yield	Reference
Alkaline phosphatase	277	35%	HARPER and ARMSTRONG (1972)
Amino-acyl-tRNA binding	24.8	3.9%	PELLEY and STAFFORD (1970)
Esterase-5	66	24%	NARISE and HUBBY (1966)
Glucose-6-PO_4 dehydrogenase	242	38%	STEELE *et al.* (1968)
Lactate dehydrogenase	130	20%	RECHSTEINER (1970a)
Octanol dehydrogenase	75	15.5%	SIEBER *et al.* (1972)
Tryptophan oxygenase	60	27%	BAILLIE and CHOVNICK (1971)
Sucrase	87	2.4%	HUBER and LEFEBVRE (1971)

fruitful. It is beyond our scope and intent to describe these procedures. However, there are some peculiarities specific to *Drosophila* (and probably related organisms) that are worthy of mention.

Special Problems

Phenol Oxidase. Anyone who has made a homogenate of *Drosophila* is well aware of the fact that if allowed to stand for any length of time, it will turn black. This is due to phenol oxidase activation which occurs subsequent to homogenization. These active phenol oxidases are then able to metabolize their substrates, which are present, resulting in quinone production and melanin formation. This phenomenon can cause problems for the enzymologist in two ways. First, and probably most serious, are the quinone products. These compounds are highly reactive. They can react with free amino groups of a protein causing denaturation. They may also interfere with assays by reacting with reagents used or products formed. Secondly, phenol oxidase from other sources and possibly from *Drosophila* has the ability to oxidize intrapeptide tyrosines. If this were to occur, denaturation of a protein is likely. Therefore the production of quinones and/or phenol oxidase activity should generally be prevented. This can be accomplished without difficulty by using one of a variety of procedures. The choice depends on the most suitable conditions for the enzyme under investigation. The substrates for phenol oxidases, being aromatic compounds, can be removed prior to quinone production by adsorption with charcoal. Many investigators have used such a procedure routinely in the preparation and assay of a variety of *Drosophila* enzymes. The use of charcoal or other adsorbants may also remove other aromatic molecules which could be inhibitory to a particular enzyme. A possible risk of this procedure is that adsorption may also remove cofactors necessary for the activity under study.

Phenol oxidase activity can also be prevented by the use of inhibitors of the enzyme. A variety of agents have been used. The most common is phenylthiourea at a concentration of 1 mM, or added as a saturated solution to 2% of saturation in homogenizing and assay solutions. Other compounds effective in inhibiting phenol oxidase are 1 mM cyanide, 40 μM diethyldithiocarbamate and 1 mM cysteine (HOROWITZ and FLING, 1955). Disulfide reducing agents, 2-mercapto-ethanol and dithiothreitol can also be used. In the study of oxidative enzymes having metal requirements, the use of these compounds may be inappropriate. Phenol oxidase is relatively heat sensitive, and thermal inactivation can be accomplished. This approach is useful in the detection of phenoxazinone synthetase activity (PHILLIPS and FORREST, 1970). Since phenol oxidase is a latent enzyme, reagents which prevent the activation process may also be useful. These include 1.5 M KCl, 0.1% sodium dodecyl sulfate and urea (MITCHELL *et al.*, 1967). With this variety of approaches available it should be possible to find an appropriate condition in which any possible effect of phenol oxidase activity on another enzyme can be eliminated. The problem is clearly of sufficient magnitude that possible effects should be routinely eliminated, or at least assessed before measuring enzyme activity or attempting enzyme purification from *Drosophila* extracts.

Protease. Another problem which is apt to be encountered using *Drosophila* is high protease activity. Highest protease activity is found during the larval stages, but significant activity is found at all stages of development (WALDNER-STIFELMEIER, 1967). The high levels of protease in whole animal extracts should be expected during the feeding stages, since the digestive tract, which has proteolysis as a main function, is included in these extracts. In addition protease, possibly of several types, must undoubtedly be involved in the extensive histolysis involved in holometabolous development which occurs in *Drosophila*. This problem has not been as generally appreciated as the phenol oxidase problem. Clearly is is not as visible to the casual observer as is a black homogenate. However, there are precedents for effects on several enzymes which are likely to be due to the action of proteases. CHRISTOPHER *et al.* (1971) noticed that the activity of phenylalanine-tRNA synthetase in larval extracts increased 50% when 1 mM phenylmethylsulphonyl fluoride, a protease inhibitor, was included in the extracts. SCHNEIDERMAN (1967) found that a larval isozyme of alkaline phosphatase could be converted to a pupal type isozyme by treatment with proteases. Replacement of the larval isozyme by the pupal form occurs normally during development. SCHNEIDERMAN'S data suggest that this replacement is due to conversion by proteases which are in the degenerating larval digestive tract (see p. 126).

Obtaining Material

Many recent advances in enzymology have come about due to the development of techniques for the mass culture of animals. For enzyme purification it is usually necessary to begin with 10 to 100 grams of animals (10^4 to 10^5 organisms). Furthermore, it is quite evident that the activity of various enzymes fluctuates during development. Therefore reasonable synchrony is often desirable, if not essential, for efficient enzyme purification. It is beyond our scope to consider the general aspects involved in the handling and maintenance of *Drosophila* cultures. An excellent recent review of these subjects has been compiled by DOANE (1967). We would like to point out the potential offered by mass culture techniques for preparative enzymology. A procedure for mass culture was developed by MITCHELL and MITCHELL (1964), and has subsequently been used by several workers. The basis for the technique is the collection of eggs over a short period of time from a large breeding population, usually on the order of several tens of thousand flies. It is quite simple to collect on the order of 10^4 to 10^5 (10–100 grams) animals per day in this way, and undoubtedly higher yields are possible. The larval synchronization is based on a short egg-collection period. Animals can be resynchronized by taking advantage of the air bubble which forms in prepupae several hours after puparium formation. Prior to air bubble formation, animals sink in water, and subsequently, they float. After synchronization animals can be allowed to develop until the desired stage. The essential point to recognize is that large numbers of reasonably synchronous animals can be readily obtained. This makes most enzymological approaches entirely feasible.

Comparing Enzyme Activities

Specific conceptual problems must be faced if comparisons of enzymes between strains are to be meaningfully interpreted. It is common practice to measure enzyme levels in a mutant and compare this measurement with a wild-type strain. Differences are usually interpreted as indicating a relation between that gene and the enzyme under study. However, the facts of development must be taken into account if these comparisons are to be valid. A casual survey of the literature regarding enzyme changes during development in *Drosophila* should be enough to convince one that enzyme activities are constantly changing, sometimes quite precipitously. If comparison between strains are to be valid, then measurements should always be made using animals of the same age and populations of similar synchrony. Failure to do this can either introduce artifactual differences, or obscure real differences. Ideally, the developmental profile of activity for the enzyme under study should be determined prior to strain comparisons. If possible, ages should be selected for comparison during which enzyme activities are maximal, but not undergoing marked fluctuations. Comparisons between strains using animals of mixed ages, when no developmental profile of enzyme activity has been determined, should be accepted with extreme caution.

Enzyme activities during development are often expressed as specific activity, i.e., units of enzyme activity per unit protein, or on a per live weight, or on a per organism basis. DEWHURST *et al.* (1970) have discussed the issues involved in choosing between these alternatives. They have shown that there is no difference in the precision of measurement obtainable using these three modes of expression. However, the three modes are not conceptually equivalent. It is clear that levels of enzyme activity can be affected by many factors in addition to transcription of the gene(s) which code for the enzyme. Nonetheless, the measurement of enzyme activity is usually the first step taken in assessing the activity of a genetic site during development. If this is the desired information, it is most appropriate that enzyme activity be expressed on a per organism basis. The information obtained is a reflection of the total genetic activity. Fluctuations in total genetic activity per organism can be brought about in two ways. The rate of activity of a constant number of loci can change, or conversely the number of active loci may change with the rate per locus remaining constant. A combination of these is also possible. A distinction between these possibilities can be arrived at only by measuring the enzyme activity per active genetic site. Assuming all sites per cell to be equally active, then the distinction could be approximated by expressing enzyme activity per unit of DNA in those cells where the enzyme is found. In practice this is a very difficult quantity to obtain. Using the specific enzyme activity per unit protein as a method of expression is not particularly useful for these purposes. Very significant changes occur in protein content during *Drosophila* development. These changes are independent of any particular enzyme that may be under study. Therefore, using units of protein as a basis for expressing enzyme activity during development could obscure the level of genetic activity for that enzyme. However, there are situations when assessments of enzymatic activity per unit of protein or per unit of live weight are quite useful.

If the enzyme activity of two different strains is being compared, a difference in size of the animals could result in a difference in measured enzyme level. If genetic conclusions about the particular enzyme in the compared strains is desired it would be undesirable for the conclusions to be affected by animal size, and then comparisons of specific activity at a particular stage of development would be most appropriate. Another example when specific activity measurements are useful is when one is trying to draw conclusions relative to the physiological role of an enzyme. The period during development when the activity of a specific enzyme is highest, when compared to all other protein, may indicate that this enzyme is especially important during that period. In fact, when making enzyme measurements, it is often quite useful to have measurements of total enzyme activity per animal and specific enzyme activity. These measurements can be easily performed in a single experiment.

Tissue Comparisons

A common concern of developmental studies is the measurement of relative enzyme levels in various tissues. Enzyme studies in *Drosophila* are often conducted on whole organisms, since only limited quantities of specific organs can be obtained by dissection. If a particular assay is sensitive enough to be performed on a small amount of dissected tissue, this may pose no problem (see p. 4). Often assay procedures can be adapted to a micro scale. However, several investigators have recently described procedures for the bulk isolation of specific organs. These may become quite useful for studying tissue localizations, or for purification of enzymes from particular tissues. KNOWLES and FRISTROM (1967) developed a procedure for isolating gram quantities of integument. Imaginal discs were isolated in bulk by FRISTROM and MITCHELL (1965). Salivary glands from *D. hydei* have been isolated by BOYD *et al.* (1968), and from *D. melanogaster* by COHEN and GOTCHEL (1971). The most elaborate bulk isolation procedure is the one developed by ZWEIDLER and COHEN (1971). This procedure can yield large amounts of salivary glands, imaginal discs, testes, Malpighian tubules, gut and fat bodies. These procedures have several aspects in common. A special method is developed for tearing open the larvae. The larvae are squeezed to expel the organs, and then the organs are settled and decanted in various ways. Finally density gradient centrifugation through ficoll gradients is used to obtain the pure fractions. Using these procedures it appears feasible to isolate separate organs in gram quantities.

Another approach to the determination of tissue distributions has recently been used, namely, histochemical procedures as they are used in electrophoretic gel assays. In these assays a colored, insoluble product is formed at the site of enzyme activity in the tissue, in a manner analogous to the formation of a band in a gel. In this way, tissue localization of aldehyde oxidase has been determined by DICKINSON (1971, see also p. 66), lactate dehydrogenase and α-glycerophosphate dehydrogenase by RECHSTEINER (1970, see also p. 89), octanol dehydrogenase by MADHAVAN *et al.* (1973, see also p. 79). This approach has the advantage of being able to distinguish areas within a tissue that have activity from those that do not. This is superior to homogenization of the whole organ, that results in obtaining the average activity throughout the entire organ. A disadvantage is that it is a descriptive procedure and can only be semi-quanitative.

Eye Color Mutants and Their Enzymes

Ommochrome Synthesis

Introduction

The eye color mutants of *Drosophila* occupy a significant place in the history of genetics in general, and biochemical gentics in particular. The first mutant to be described in *Drosophila* was *white* eye *(w)* by T. H. MORGAN (MORGAN, 1910). This observation marked the beginning of the extensive work of MORGAN, his co-worker BRIDGES *et al.* and their students, which developed the genetic analysis of this small insect to the degree of sophistication it has reached today. The experiments of BEADLE *et al.* during the 1930's and early 1940's on the development of eye colors and the interaction of genes responsible for eye color production laid much of the groundwork for our knowledge of the relation between genes and metabolism. These were some of the earliest experiments that linked genes to a sequence of steps in a metabolic pathway. This work has been reviewed extensively (EPHRUSSI, 1942; BEADLE and TATUM, 1941).

There are two classes of pigments in the eye of *Drosophila*. The brown pigments, called ommochromes, are products of tryptophan metabolism. The other class, usually referred to as the red pigments, are pteridines which are products of purine metabolism. The genetic aspects of the production of these pigments in *Drosophila* and other insects has been reviewed by ZIEGLER (1961).

Analysis by Transplantation

EPHRUSSI and BEADLE (1936) developed the eye disc transplantation technique in order to study the role of various genes and their interaction in controlling the development of the normal reddish brown color which is characteristic of the wild-type eye. In essence, this technique consists of dissecting out an eye imaginal disc from a larva of desired age, and implanting it into the body cavity of an appropriate host larva. The host is allowed to proceed through metamorphosis, and the abdomen of the resulting adult is then examined to determine the fate of the now differentiated eye. BEADLE and EPHRUSSI (1936) found that when eye discs from various mutants were cultured in wild-type hosts, they would develop the mutant eye color according to their own genotype. However, there were two exceptional ommochrome deficient mutants, *vermilion (v)* and *cinnabar (cn)*. These mutants are characterized by a lighter, brighter red eye. When cultured in wildtype hosts, they did not develop according to their own genotype, but behaved nonautonomously and had wild-type pigmentation. These experiments led BEADLE and EPHRUSSI to postulate the existence of substances called the v^+ substance and the cn^+ substance. The production of these substances is a function of the wild type alleles of the *v* gene and *cn* gene respectively, and they are necessary for the development of normal pigmentation. The non-autonomy of *vermilion* and *cinnabar* was explained by assuming that the wild-type environment supplied these substances to the transplanted discs, thereby allowing the development of normal eye colors. The results obtained from reciprocal transplants between *vermilion* and *cinnabar* allowed for the determination of the order in which the v^+ substance and the cn^+ substance functioned. *Vermilion* discs implanted into a *cinnabar* host resulted in a transplant with wild-type eye pigment. However, *cinnabar* discs implanted into a *vermilion* host resulted in a transplant having the mutant phenotype. The conclusion was that genes v^+ and cn^+ functioned in sequence, and that the v^+ gene must function before the cn^+ gene, since in the *cn* host there was enough v^+ substance to allow a *vermilion* disc to form wild-type pigments. However, in the *vermilion* host neither v^+ or cn^+ substances produced. Therefore the implanted *cinnabar* disc remained unable to make any brown pigment.

Identification of Enzymes

The association of the eye-color mutants with specific enzymes was made possible by several advances occurring after the work on disc transplants. TATUM and HAAGEN-SMIT (1941), showed that the v^+ substance was kynurenine, a metabolic derivative of tryptophan. Some years later the cn^+ substance was identified as 3-OH-kynurenine by BUTENANDT *et al.* (1956). The structural chemistry of the brown-eye pigments was determined primarily by the extensive work of BUTENANDT and his co-workers. The chemistry of ommochromes has been the subject of recent reviews (BUTENANDT and SCHAFER, 1962; FORREST, 1959; LINZEN, 1967). The brown pigment in the eye of *Drosophila* has been shown to be xanthommatin. Xanthommatin is a member of the class of ommochromes called ommatins. It is a phenoxazone dye, and may be a precursor to the other ommatins and the more complex, high molecular weight ommins. The sequence of enzymatic reactions leading to xanthommatin is shown in Fig. 10.

Enzyme activities which will catalyze each step in this sequence have been demonstrated in *Drosophila*. Two mutants, *v* and *cn*, have been associated with specific enzyme deficiencies which account for the mutant phenotype. However, it is not definitely known whether this pathway has any other physiological significance in addition to eye pigmentation. There are several mutants which cause reduction or absence of ommochromes, but cannot be associated with any specific enzymatic step. Therefore, there are many questions that remain to be answered concerning genes affecting eye pigmentation and specific enzymes. It is to be hoped that as more detailed biochemical information becomes available, these questions will be answered.

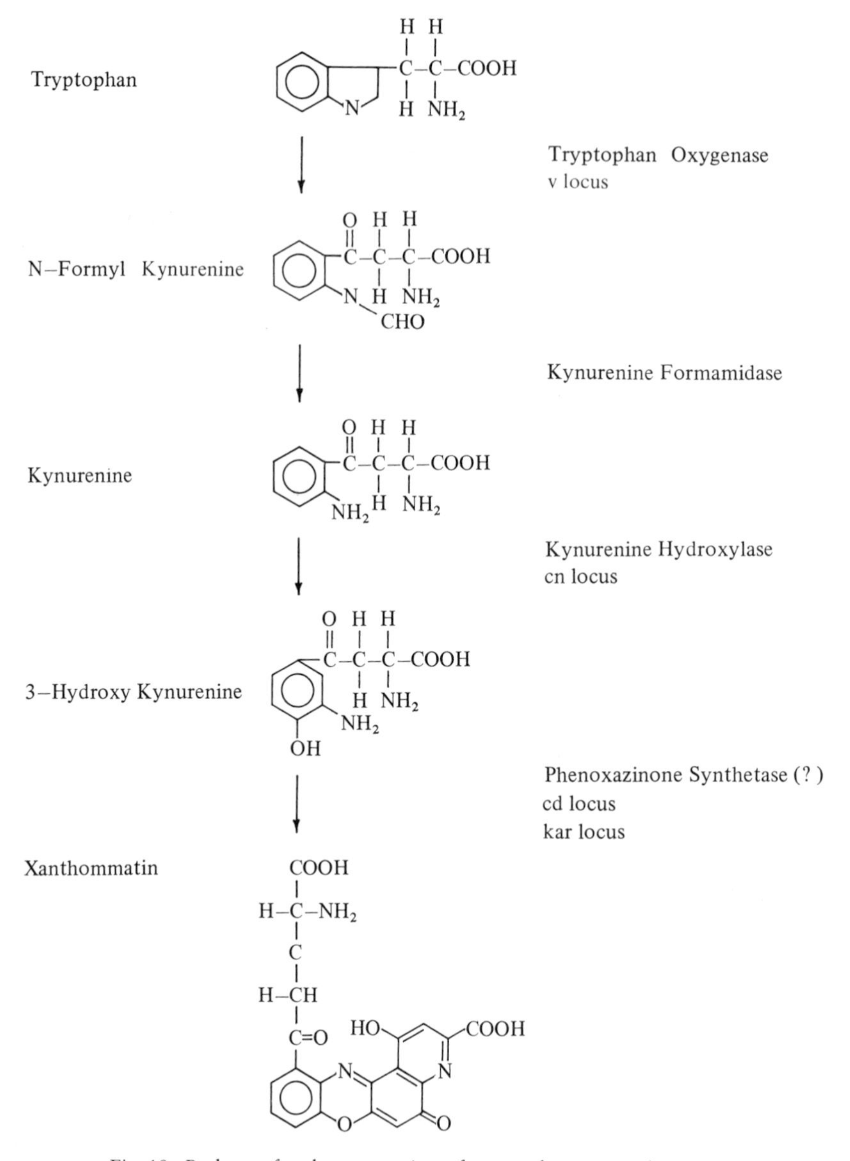

Fig. 10. Pathway for the conversion of tryptophan to xanthommatin

Tryptophan Oxygenase (Pyrrolase) (EC 1.13.1.12)

Biochemistry of Tryptophan Oxygenase

The first enzyme in the ommochrome pathway is tryptophan oxygenase. This has been the most extensively studied enzyme in the ommochrome synthetic pathway. Its activity was initially described by BAGLIONI (1959). A number of investigators have subsequently studied this enzyme; KAUFMAN (1962), MARZLUF (1965a), and TARTOF (1969). However, appreciable enzymological information has only recently become available (BAILLIE and CHOVNICK, 1971). For the detection of activity it is essential that extracts be maintained with a reducing agent. Ascorbate was used by KAUFMAN (1962), and recently 2-mercaptoethanol has been found more satisfactory (MARZLUF, 1965a; BAILLIE and CHOVNICK, 1971). The enzyme is dialysis labile and is inhibited by hydroxylamine, azide and by Cu^{++}. There is partial inhibition by sodium diethyldithiocarbamate, D-tryptophan and 5-methyltryptophan (MARZLUF, 1965a). TARTOF (1969) has found the pH optimum for tryptophan oxygenase from wild-type *Drosophila* to be 7.4. He has also found the apparent K_m for tryptophan to be 1.5×10^{-3} M.

The most extensive biochemical investigation of tryptophan oxygenase has recently been reported by BAILLIE and CHOVNICK (1971). These authors have reported several procedures which assure that maximal levels of activity can be obtained in extracts and partially purified preparations. They have observed that the inclusion of neutral Norite during homogenization results in a more than two-fold increase in measured activity. This may be due to the absorption of potential inhibitions such as pteridines which are present in *Drosophila* extracts, and are known to inhibit tryptophan oxagenase (GHOSH and FORREST, 1967a). Suspecting that this enzyme might have a heme co-factor, MARZLUF (1965a) tried to stimulate activity by adding hematin, but without success. However, BAILLIE and CHOVNICK find that the inclusion of methemoglobin, presumably to act as a hematin donor, gives a significant stimulation. The magnitude of this stimulation is around 2-fold in crude extracts, and increases with purification. In samples that are approximately 60-fold purified, obtained by combining ammonium sulfate fractionation, gel filtration and ion exchange chromatography, activity is almost entirely dependent on added methemoglobin. The recovery in this 60-fold purified preparation was 27%. BAILLIE and CHOVNICK (1971) have also reported that there is a lag of about 30 minutes in the enzyme assay. This can be overcome by preincubating the enzyme in methyltryptophan, but not tryptophan. The molecular weight of *Drosophila* tryptophan oxygenase, estimated by gel filtration and velocity sedimentation in sucrose gradients, was found to be in the range of 150000.

Another consideration that should be taken into account in evaluating studies on tryptophan oxygenase is the fact that the immediate product of the tryptophan oxygenase catalyzed reaction is formylkynurenine, while kynurenine production is measured in the assay. The conversion of formylkynurenine to kynurenine is accomplished by another enzyme, kynurenine formamidase. GLASSMAN (1956) and KIMMEL (1969) have shown that high levels of the formamidase are found in *Drosophila*. It is usually assumed that the formamidase is in excess, however this assumption does not always seem resonable. It implies that kynurenine formamidase is present in amounts sufficient to convert all the formylkynurenine produced by tryptophan oxygenase in assays of all tissues, at all stages of development and more importantly during all steps of tryptophan oxygenase purification. A more reasonable approach is to use procedures that will assure complete conversion of formyl kynurenine to kynurenine when assaying for tryptophan oxygenase. There are several approaches to this problem. The simplest is that used by BAILLIE and CHOVNICK (1971). It is known that formylkynurenine breaks down to kynurenine in dilute acid at 90° C. After stopping the enzyme reaction with 5% TCA, BAILLIE and CHOVNICK simply heated all assays at 90° for 10 min. It should be noted that these precautions were not always followed in the developmental studies reported below.

Genetics of Tryptophan Oxygenase

The mutant *vermilion* has been shown to lack the ability to produce kynurenine. This has been attributed to the absence of detectable tryptophan oxygenase (BAGLIONI, 1959; KAUFMAN, 1962; TARTOF, 1969). The *vermilion* locus (1.33.0) has been analyzed structurally by GREEN and found to be subdividable genetically and physiologically (GREEN, 1952, 1954).

The allele v^1 and v^{36F} are separable by recombination. Wild-type flies, demonstrated to be recombinants using outside markers, are obtained in a cross between flies that are v^1 and v^{36F} respectively. Females which are $\frac{+\,v^{36F}}{v^1 +}$ are mutants, whereas $\frac{+\;+}{v^1\,v^{36F}}$ are wild-type. This failure of these and other v alleles to complement indicates that although they are separable by recombination, the v locus apparently is a single functional unit.

The v alleles also can be differentiated by their response to a starvation diet, and to a suppressor mutation, *suppressor of sable su-(s)*. When *vermilion* larvae of an appropriate age are placed on a starvation diet, they develop into flies that show partial restoration of brown pigment (BEADLE *et al.*, 1938). SHAPARD (1960) and GREEN (1954) showed that only those v alleles which were suppressible by *su (s)*, referred to as v^s alleles, responded to starvation in this manner. The v alleles which are not suppressible, v^u alleles, do not produce pigment upon starvation. SHELTON *et. al.* (1967) have shown that this partial restoration of eye color is due to an increase in the tryptophan oxygenase levels which can be detected in pupae raised on the starvation diet. Several investigators, BAGLIONI (1960), KAUFMAN (1962), MARZLUFF (1965 b) and TARTOF (1969) have shown that suppression of the v gene results in the increase in tryptophan oxygenase levels. Various models for the v locus are possible and have been proposed, e.g. MARZLUF (1965 b). These postulate that one or the other class of v alleles or *su (s)* would be mutations in some part of a regulatory system. While the genetics of this locus and the biochemistry of its enzyme, tryptophan oxygenase, are far from complete and therefore various models are possible, there does not seem to be any compelling evidence to rule out the most simple hypothesis, namely, that the various v alleles all represent different mutations of the structural gene for tryptophan oxygenase. Recent evidence on the nature and action of *su (s)* is not compatible with its being part of a simple regulatory system for tryptophan oxygenase (see below).

Evidence that v is a structural locus for tryptophan oxygenase has recently been provided by the work of TARTOF (1969) and BAILLIE and CHOVNICK (1971). The essential feature of this evidence involves the demonstration that in the mutant, v^k, an altered protein is made. The properties of this molecule are consistent with its having an altered primary structure. TARTOF (1969) has shown that the suppressible v alleles make a product that will complement with wild-type extracts to give more than additive levels of activity. The factor in one of these, v^k, responsible for this additive effect is heat labile, nondialyzable and can be precipitated with ammonium sulfate in a similar manner as native tryptophan oxygenase, and presumably is a protein. BAILLIE and CHOVNICK have shown that this v^k product chromatographs on Sephadex G-200 at the same position as native tryptophan oxygenase, and therefore is of approximately similar size.

The pH optimum for the tryptophan oxygenase obtained from suppressed v^k flies is shifted relative to the normal enzyme, and the apparent K_m for substrate is different. These alterations in catalytic activity are not dependent on the nature of the suppressing allele, since the enzyme from v^k is similar in the presence of several different alleles of *su (s)*, su^2 *(s)*, su^3 *(s)* and su^{51C15} *(s)*. None of these suppressor alleles has any effect on the pH optimum or K_m of wild-type enzyme. When v^1 is suppressed, the enzyme which is produced is similar to wild-type with respect to pH optimum, K_m, and thermal stability. Therefore, it appears that the altered properties of v^k enzyme are due to the v^k gene itself (TARTOF, 1969).

The suppression of v mutants is the only case in *Drosophila* where suppression of a gene associated with a specific enzyme has been studied. MARZLUF (1965b) investigated the properties of tryptophan oxygenase produced in a suppressed *vermilion* fly. He observed that the suppressed v enzyme was similar in several respects to wild-type. From this he concluded that suppression permits the production of a small quantity of normal enzyme. Recently, TARTOF (1969) has conducted a more extensive series of investigations, in which he examined properties of tryptophan oxygenase from several *vermilion* alleles in combination with three different alleles of *su (s)*. In order of decreasing enzyme activity when suppressed, the v alleles are arranged as follows: v^k, v^1, v^{36F}. The alleles v^{48a} and v^{51c} do not have any tryptophan oxygenase activity and are classified as unsuppressible. The order of these v alleles is independent of which *su (s)* allele is used to suppress them. The absolute amount of enzyme activity measured is also independent of the suppressing allele. TARTOF did not observe any qualitative effect on enzyme properties specific to any of the three *su (s)* alleles; su^2 *(s)*, su^3 *(s)* or su^{51c15} *(s)*. The *su (s)* gene appears to be completely recessive in its ability to restore enzyme activity to v flies. TARTOF argues that for several reasons *su (s)* cannot be an informational suppressor that acts at the transcriptional or translational level. It is already known that *su (s)* is not allelic to v, and therefore cannot be an intragenic suppressor. TARTOF therefore concludes that *su (s)* acts at the metabolic or physiological level to create an environment in which enzyme activity can be restored to the mutant protein. This conclusion implies that suppressible mutant v genes make a product that for some reason is inactive. Evidence for this inactive product has been provided by TARTOF (1969).

TARTOF has observed that the suppressible v mutants are not entirely recessive. Heterozygotes of v^+ and various mutant v alleles were examined for their tryptophan oxygenase levels. It was observed that the heterozygotes showed greater than 50% of the wild-type levels. Furthermore, the amount of enzyme activity in heterozygotes of wild-type and various v alleles could be ranked in an order that is identical to the order of enzyme production, when the v alleles are suppressed, i.e. v^k, v^1, v^{36F}, v^{51c}, v^{48a}. This observation was carried one step further when TARTOF found that *in vitro* mixtures of extracts of wild-type and suppressible v mutants also showed an increased enzyme activity over wild-type alone. The magnitude of the increase obtained when equal amounts of wild-type and mutant extracts were mixed, closely approximates the levels seen in the corresponding heterozygotes. The factor in the v^k mutant has properties (see above) that are similar to native tryptophan oxygenase. It therefore appears that the suppressible *v* mutants make a form of the enzyme which is normally not active, but can be activated by a change in the environment. This change can be brought about by the gene *su (s)*, starvation, or by putting the mutant protein in the presence of wild-type enzyme.

Evidence as to the nature of the events which occur upon suppression has recently been provided by TWARDZIK *et al.* (1971). These authors set out to examine the possible role of transfer RNA in mediating suppression by the *su (s)* gene. Chromatographic separation and analysis of *Drosophila* tRNA species show that there is a difference between wild-type and *su (s)* in the tRNA species which will accept tyrosine. The *su (s)* flies show a virtual absence of one of the isoaccepting forms of tyrosyl-tRNA, and an excess of another tyrosyl-tRNA, but the total amount of tyrosyl-t-RNA in wild-type is the same as in *su (s)*. The following genetic experiments confirm that *su (s)* is the gene responsible for the absence of the specific tyrosyl-tRNA. Appropriate crosses show that the gene responsible for the alteration in tRNA pattern is X-linked and recessive, as is *su (s)*. An independent allele of *su (s)*, referred to as $su\ (s)^{e1}$, was induced by EMS mutagenesis. This new allele also causes the absence of the same isoaccepting form of tryosyl-tRNA, which is absent in $su\,(s)^2$ flies. TWARDZIK *et al.* (1971) concluded that since *su (s)* is recessive and affects the distribution of tRNA species and not the absolute amount of tRNA present, the wild-type allele of *su (s)* is a gene which functions to produce a factor involved in tRNA modification, and is not a structural gene for tyrosyl-tRNA. The relation between the absence of a particular tRNA and the restoration of tryptophan oxygenase accomplished by *su (s)* has been examined by recent experiments of JACOBSON (1971). He found that extracts from *vermilion* animals could be activated, *in vitro*, by treatment with ribonuclease to produce tryptophan oxygenase activities comparable to wild-type. Furthermore, this activated tryptophan oxygenase could be specifically inhibited by the addition of a specific isoacceptor form of tyrosyl-tRNA. This is the same species of tRNA which is lacking in the homozygous *su (s)* animals. The mechanism of suppression that emerges from these experiments is that in suppressible *v* mutants an altered form of the enzyme is produced. This mutational alteration makes the enzyme succeptible to inhibition by a specific isoacceptor form of tyrosine tRNA which is normally present in wild-type flies. Suppression is accomplished by a second mutation, in *su(s)*, which results in the absence of this inhibitory tRNA. The altered tryptophan oxygenase is then functional. JACOBSON assumes that the specific tyrosyl-tRNA is usually associated with tryptophan oxygenase in wild-type flies. Verification of this assumption would be most interesting, and might point to an important regulatory function. However, at present the possibility exists that the inhibition of the altered tryptophan oxygenase by the tyrosyl-tRNA is just a spurious interaction between two substances that are found in the same cytoplasm. At present there is no obvious physiological connection between tyrosyl-tRNA and the conversion of tryptophan to eye-pigments.

Several other genetic aspects of tryptophan oxygenase which pertain to its regulation have been investigated. RIZKI and RIZKI (1963) have reported that animals fed on a high tryptophan diet have a higher level of kynurenine in their fat bodies and has suggested that the enzyme tryptophan oxygenase may be inducible. The mechanism by which this increased activity is brought about is not known and might warrant further investigation.

TOBLER *et al.* (1971) and BAILLIE and CHOVNICK (1971) have used the X-linked enzyme tryptophan oxygenase to study dosage compensation. As noted previously by KAUFMAN (1962), tryptophan oxygenase is dosage compensated, i.e. males and females have equal levels of activity, even though females have two doses of the v^+ gene and males only one. These levels are not due to a stimulating factor of the Y chromosome, because X/O males show levels comparable to X/Y males. Translocations of the v^+ gene to the third chromosome result in a lowering of activity for some unknown reason, but the levels of activity in males and females with this translocation are approximately equal, showing that X location itself is not a condition for dosage compensation. Also v^+ genes translocated to the Y chromosome are dosage compensated. Males which carry v^+ on the X and an additional dose on the Y show the activity expected by simply adding the level observed in normal v^+ males to that observed in males carrying their only v^+ allele translocated to the Y. The general conclusions reached by TOBLER *et al.* are that a v^+ allele has a unique level of activity which is dependent on its location in the genome, and these levels are simply additive. However, the dosage compensating mechanism is not dependent on location. Enzyme activity per gene dose in males is always twice that of females, regardless of where th v^+ genes are located. BAILLIE and CHOVNICK (1971) reach the same conclusion. These studies as yet have not given an explanation of the mechanism of dosage compensation. However, a well characterized gene-enzyme system, such as the *v* locus and tryptophan oxygenase should be a most fruitful place to conduct informative experiments on the problem of dosage compensation and its mechanism.

A potentially interesting study was conducted by MORRISON and FRAJOLA (1964), who report that tryptophan oxygenase can be synthesized in an *in vitro* protein synthesizing system. Activity was dependent on a fraction from wild-type, but similar fractions from *vermilion* animals showed no activity. However, PHILLIPS *et al.* (1967) have not been able to confirm these results, and claim the apparent enzyme activity was due to a non-enzymatic reaction between tryptophan and ascorbic acid which was used in their assay.

Developmental Biology of Tryptophan Oxygenase

Developmental characterization of tryptophan oxygenase has not been extensive. KAUFMAN (1962) has determined the tryptophan oxygenase activity of whole animals throughout development. Activity gradually increases during the larval stages, remains fairly constant during pupal stages, and increases sharply, about two-fold, after emergence. Activity shows a gradual decline beginning around the fourth day of adult life.

KAUFMAN has shown that activity is distributed between the head (40%), thorax (40%) and abdomen (20%) of adult flies. RIZKI (1963) has claimed that all the activity in late third instar larvae is found in the fat body. RIZKI (1961) has also studied the cytological distribution of kynurenine, a secondary product of the enzyme reaction, and finds it is confined to the anterior part of the larval fat body.

A variation of the transplantation technique of BEADLE and EPHRUSSI has been used to detect the presence of the v^+ substance, kynurenine, in various tissues. A suspected source of kynurenine is implanted into a test host. A convenient test host is provided by the double mutant *white-apricot* (w^a), *vermilion* (v). This host can make little if any kynurenine, and has very light eyes. An increase in the amount of kynureine will lead to a darkening of the host eyes. This procedure has been used to detect the release of kynurenine from various tissues. Malpighian tubules, the fat body of the late third instar larvae, and the eye discs obtained from late third star larvae have been shown to release kynurenine, (EPHRUSSI, 1942). Wild-type eye discs implanted into *vermilion* hosts develop into eyes with wild-type pigmentation, which also indicates the ability of developing eyes to produce kynurenine. The above experiments only measure the presence of releasable kynurenine, and do not directly measure synthesis. These results can be explained by assuming that the tissues indicated have the ability to remove kynurenine from the hemolymph, store it, and subsequently release it when transplanted. It has been shown that Malphigian tubules do have the ability to pick up and store kynurenine (BEADLE, 1937a). However, the v^+ substance (kynurenine) can be detected in the tissues before it can be detected in the hemolymph (BEADLE, 1937b). Until precise measurements of enzyme activity in each tissue during development are available, the situation will remain unclear. The most satisfactory hypothesis at this time would predict that several tissues, fat body, possibly Malphigian tubules, the developing eye, and others (?) have the enzymatic capabilities to synthesize kynurenine. Probably more kynurenine is synthesized than will be used in eye pigmentation in wild-type flies. The developing eye apparently has some flexibility as to where it will get sufficient kynurenine to produce the wild-type phenotype. It may synthesize it, or get it from the hemolymph.

Kynurenine Formamidase

The second enzyme in the pathway of ommochrome biosynthesis is kynurenine formamidase. This enzyme is responsible for the conversion of formylkynurenine to kynurenine. The activity has been described by GLASSMAN (1956). Little is known concerning the biochemistry of this enzyme. It is inhibited by bisulfate. It does not use formylanthranylate as a substrate. Normal levels of formamidase activity have been demonstrated in many eye color mutants (GLASSMAN, 1956). At present no connection has been established between this enzyme and a specific gene.

The developmental biology of this enzyme has not been studied directly. However, kynurenine production is involved in the assay of tryptophan oxygenase, and therefore one might reasonably expect to find this enzyme in the same tissues that one finds tryptophan oxygenase. Preliminary information (KIMMEL, 1969, 1970), indicates this enzyme may be present throughout *Drosophila* development.

Kynurenine Hydroxylase (EC 1.14.1.2)

Biochemistry of Kynurenine Hydroxylase

The third enzyme in this pathway is kynurenine hydroxylase. This enzyme catalyses the conversion of kynurenine to 3-hydroxykynurenine. The enzyme from *Drosophila* was initially described by GHOSH and FORREST (1967b). They observed that the activity is found in a pellet fraction obtained by centrifuging the supernatant of a 1000×g, 10 min centrifugation at 37000×g for $1^1/_2$ hrs. They suggested it might be a mitochondrial enzyme. In other systems, mammals and fungi, this is the case. Its mitochondrial localization in *Drosophila* has recently been verified by SULLIVAN *et al.* (1973), who showed that kynurenine hydroxylase activity is found associated with particles which behave like mitochondria in a differential centrifugation experiment. The particles sediment in velocity and equilibrium sucrose gradients in a manner which is identical to the particles which contain cytochrome oxidase, a known mitochondrial enzyme. Electron microscopic examination of the fraction containing kynurenine hydroxylase activity revealed that it is composed primarily of mitochondria. Kynurenine hydroxylase is probably not located in pigment granules, since when the structure of pigment granules is changing i.e., during development of the eye, the sedimentation properties of kynurenine hydroxylase do not change. Furthermore the mutants *scarlet* and *white* do not have normal pigment granules (SHOUP, 1966), but the sedimentation profile of kynurenine hydroxylase containing particles is identical to wild-type.

The mitochondrial bound kynurenine hydroxylase has been studied with respect to its catalytic properties. The enzyme is specific for NADPH (GHOSH and FORREST, 1967b). It has a pH optimum of 8.1. For maximal activities, cyanide must be included in the assay, and 2-mercaptoethanol must be included during preparation or assay. The enzyme is non-specifically inhibited by a variety of metals when added in high concentrations. It is not particularly sensitive to chelating agents, however, diethyldithiocarbamate shows appreciable inhibition at 10^{-2} M (SULLIVAN *et al.*, 1973).

Genetics of Kynurenine Hydroxylase

Since this enzyme is responsible for the production of 3-hydroxykynurenine, the cn^+ substance, the prediction has been that the mutant *cn* would lack this enzyme. This prediction was verified by GHOSH and FORREST (1967b).

Other alleles of *cn* (cn^3 and cn^{35k}) also lack detectable activity. The relation between kynurenine hydroxylase and the number of cn^+ alleles has been studied by SULLIVAN *et al.* (1973). In all cases, enzyme activity is proportional to the number of cn^+ alleles present. Three dose flies were studied by using the translocation T(Y,2)C, which has a region of the second chromosome including cn^+, translocated to the Y chromosome. The activity of the cn^+ locus when on the Y chromosome is no different than in its normal position. Heterozygotes of the wild-type allele with any of the *cn* mutant alleles show one half the wild-type level. There is no indication of partial function of these mutant alleles. All of the data presently available are consistent with the hypothesis that the *cn* locus is the structural gene for kynurenine hydroxylase.

Table 3. Kynurenine hydroxylase activity in pupae with zero, one, two and three doses of the cn^+ locus

Group	Genotype	Dose of cn^+	Activity [a]
A	$+/+$	2	78.0 ± 4.81 (6)
B	$+/cn$	1	39.6 ± 2.20 (6)
C	$cn/+$	1	38.0 ± 2.84 (6)
D	$cn^{35k}/+$	1	36.5 ± 2.21 (6)
E	$cn^3/+$	1	37.2 ± 4.11 (5)
F	cn^3/cn^3, Y^{cn^+}	1	39.3 ± 3.14 (4)
G	z/z, Y^{cn^+}	3	107.3 ± 2.38 (5)
H	cn^3/cn^3	0	< 0.5 (4)

[a] Activity = mμ-moles 3-hydroxykynurenine produced per hour per 100 mg animals. Standard error of the mean indicated. The number in parentheses is the number of samples of animals assayed. Each assay was performed in duplicate.
Y^{cn^+} is Y chromosome carrying T(Y:2)C.

GHOSH and FORREST (1967b) reported that the mutant *w* had greatly reduced levels of activity. They used this observation to speculate that the biochemical basis of the *white* phenotype involved an alteration of pteridine metabolism, such that both the pteridine pigments and a proposed pteridine co-factor for kynurenine hydroxylase were missing. This would result in a block in both pathways. A re-examination of these observations by SULLIVAN *et al.* (1973) has failed to confirm their results. The mutant *w* and its alleles, w^a and w^h, have levels of kynurenine hydroxylase that do not appreciably differ from wild-type. Furthermore, there is as yet no evidence in any system which indicates a pteridine co-factor is involved in the hydroxylation of kynurenine.

Developmental Biology of Kynurenine Hydroxylase

GHOSH and FORREST (1967b) analyzed the specific activities of kynurenine hydroxylase during development. They found that activity appeared after puparium formation, reaches a maximum in mid pupae, and declines thereafter. SULLIVAN *et al.* (1973) have conducted a detailed analysis of the amounts of kynurenine hydroxylase per organism during development. This profile is shown in Fig. 11. Activity is found in early larvae, reaches a peak in early third instar and declines during the late larval stages. Activity rises again shortly after puparium formation. Maximum levels are attained at 72–76 hrs after puparium formation in eight-day-old pupae. Thereafter the levels fall rapidly to reach a value in emerged flies that is comparable to that found in larvae. The timing of the appearance of the enzyme activity coincides closely with the appearance of the eye pigmentation in the developing eye. This profile also corresponds closely with the developmental history of the cn^+ substance (HARNLY and EPHRUSSI, 1937).

The tissue distribution of kynurenine hydroxylase in pupae has been determined by SULLIVAN *et al.* (1973). Activity can only be found in the head section. More than 80% is found in the developing eye of pupae 72 hrs after puparium

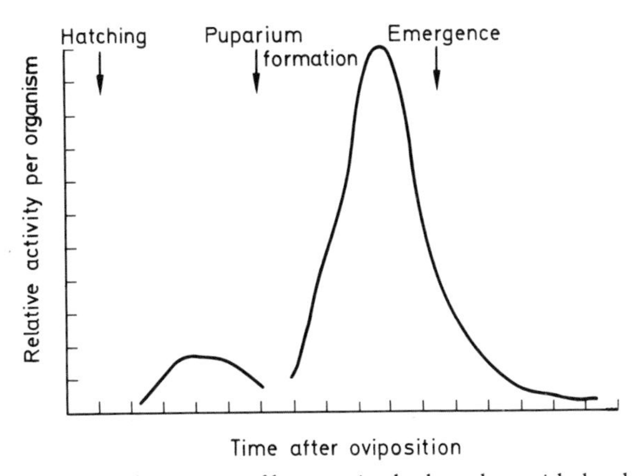

Fig. 11. Changes in the activity of kynurenine hydroxylase with development

formation, and probably all of the activity in pupae is confined to the eye, since some loss of activity occurs during dissection and assay. This indicates that kynurenine hydroxylase may be a most useful enzyme with which to study imaginal disc differentiation. It appears shortly after pupation, and it is highly specific to the product of a particular imaginal disc. The tissue localization of the larval enzyme has not been determined, however, presence of the *cn*$^+$ substance can be detected in larval Malphigian tubules, but not in the larval fat body (BEADLE, 1937c). WESSING and EICHELBERG (1968) have shown directly that 3-hydroxykynurenine is found in Malphigian tubules. However, Malphigian tubules are known to have the ability to pick up 3-hydroxykynurenine from the hemolymph (BEADLE, 1937a). Therefore it is not clearly established whether synthesis of 3-hydroxykynurenine occurs in the tubules. It could conceivably be produced elsewhere and concentrated there. Arguing against this interpretation is the failure to identify another source of 3-hydroxykynurenine, the early appearance of 3-hydroxykynurenine in the tubules relative to the eye discs, and the failure to detect 3-hydroxykynurenine in the hemolymph of larvae. Also, *vermilion* tubules can be shown to be able to produce *cn*$^+$ substance when transplanted to a *cn* host (BEADLE, 1937a). The best hypothesis at present is that kynurenine hydroxylase is present in larval Malphigian tubules.

Phenoxazinone Synthetase

The final steps in xanthommatin synthesis have remained obscure until recently. BUTENANDT *et al.* (1956) proposed that xanthommatin formation from 3-hydroxykynurenine was the result of an oxidative condensation catalyzed by dopaquinone, which was regenerated non-specifically by a diphenol oxidase system. This hypothesis remained unchallenged until called into question by PHILLIPS *et al.* (1970). They observed that mutants deficient in xanthommatin synthesis, *scarlet (st), lightoid (ltd),* and *cardinal (cd),* have normal phenol oxidase levels. This is especially noteworthy in the case of *cd*. This mutant makes little xanthommatin and accumulates 3-hydroxykynurenine, suggesting it might be expected to have a block at this step in the pathway. They also argue from the data of PEEPLES *et al.* (1969a) on *lozenge* mutants that substantial xanthommatin synthesis occurs, despite a deficiency in phenol oxidase. Therefore they questioned the proposed relation between xanthommatin synthesis and phenol oxidase.

These arguments themselves are not convincing enough to reject the hypothesis of BUTENANDT, since the phenol oxidase levels are measured in whole animals, and a deficiency in the eye might not have been detected. However, PHILLIPS and FORREST (1970) have gone on to demonstrate an enzyme activity that may be responsible for xanthommatin synthesis *in vivo*. They have called this enzyme phenoxazinone synthetase.

Biochemistry of Phenoxazinone Synthetase

Phenoxazinone synthetase catalyses the oxidative bimolecular condensation of o-aminophenols to phenoxazinones. It utilizes 3-hydroxyanthranilic acid, o-aminophenol and 3-hydroxykynurenine as substrates. It cannot use p-aminophenol, 5-hydroxy-anthranilic acid, nor kynurenine as substrates. When 3-hydroxykynurenine is used in the reaction as substrate, xanthommatin is formed. Xanthommatin formation involves, in addition to phenoxazinone formation, a quinoline ring cyclization step. This step may need a second as yet undescribed enzyme, or it may occur spontaneously. The cyclization step can occur using non-specific chemical oxidants, and PHILLIPS and FORREST do not see a decrease in this second step relative to phenoxazinone synthesis when they manipulate the enzyme activity in various ways. These facts are consistent with the second step being spontaneous. But as PHILLIPS and FORREST point out, there are several mutants which block xanthommatin formation *in vivo*. These could be interpreted to be defects in several consecutive enzymes. However, it appears that until evidence is provided to the contrary, spontaneous cyclization of the side chain of the phenoxazinone to yield xanthommatin is a satisfactory hypothesis.

This enzyme has the interesting property of undergoing activation. This can be accomplished by passing extracts made in 1.5 M KCl over a Sephadex G-50 column, or by heating in a 71° C water bath for 60 sec. The two procedures are not additive, indicating the same activity is being generated by both treatments. The phenomenon of heat activation is also useful to distinguish the phenoxazinone synthetase activity from possible competing reactions, such as cytochrome oxidase or phenol oxidase. Both of these enzymes are completely inactivated by heat treatment (PHILLIPS and FORREST, 1970). Therefore, it appears that *Drosophila* has an enzymatic activity, separate from phenol oxidase, that can function to catalyze xanthommatin synthesis.

In a recent report PHILLIPS *et al.* (1973) have observed that phenoxazinone synthetase occurs in both a soluble and particulate form. The soluble fraction is quite large, however, since it reaches its equilibrium position in a 0.2 M to 2.0 M sucrose gradient after 45 min of centrifugation at 65000 × g. The particulate fraction is found at the same position of the gradient as is a heavily pigmented band. This suggests a possible pigment granule association for the enzyme. These observations, along with the observations on several mutants, have led these authors to postulate a model for ommochrome biosynthesis in which all of the enzymes on this pathway are associated with developing pigment granules. One feature of this model is that normal levels of all enzymes are necessary for the integration of the enzymes into a multi-enzyme complex. This model has many attractive features, and would be useful in the interpretation of several of the existing mutants in the ommochrome pathway whose biochemical bases are not understood. However SULLIVAN *et al.* (1974), have shown that tryptophan oxygenase and kynurenine formamidase are soluble enzymes, while kynurenine hydroxylase is a mitochondrial enzyme. There is no evidence for the association of the first three enzymes of the pathway with multi-enzyme complexes or pigment granules, and this model appears unsuitable.

Genetics of Phenoxazinone Synthetase

There is as yet no direct evidence as to the genetic locus for this enzyme. However, there is circumstantial evidence that the *cardinal (cd)* locus may be a structural gene for phenoxazinone synthetase. PHILLIPS *et al.* (1970) observed that its substrate, 3-hydroxykynurenine, accumulates in *cd* at the time pigment should be synthesized. PHILLIPS *et al.* (1973) measured phenoxazine synthetase levels in heads of adult flies of several genotypes, and reported that the levels of this enzyme in *cd* were approximately 40% of wild-type. They feel this level is compatible with the "leaky" phenotype of *cd*. The *cd* mutant only makes 15% of the wild-type level of ommochromes. While it is difficult to judge the phenotypic effects of partial enzyme function, it appears that 40% of wild-type enzyme levels may be higher than one might expect for this level of pigment synthesis.

PHILLIPS *et al.* (1973) have also observed that the levels of phenoxazinone synthetase in other mutants defective in ommochrome synthesis are also much lower than wild-type. These include 8.6% in *lightoid (ltd)*, 11.1% in *st*, 12.6% in *v*, 16.7% in *cn*, and 29.6% in *w*. These low levels serve as another basis for the multi-enzyme complex model discussed above. With the exception of *cn*, none of these mutants affect the level of kynurenine hydroxylase activity (SULLIVAN *et al.*, 1973).

Pteridine Pigments

Introduction

The investigations of the pteridine based red eye pigments in *Drosophila* have been less successful than the corresponding work on the ommochromes in determining specific synthetic pathways, and uncovering the enzymatic basis for various mutants. However, an examination of this work does provide some interesting insights into the problems and frustrations of relating even the simplest visible phenotypes to specific enzymatic defects. Despite over twenty years of intensive work by several groups, the metabolic pathways involved remain unclear, and the enzymatic lesion associated with altered pteridine content has been identified in only two closely related cases. Reviews can be found in ZIEGLER (1961), ZIEGLER and HARMSEN (1969), and HADORN (1958).

The pteridines of biological interest are primarily the 2-amino-4-hydroxy derivatives of the pteridine ring. This family of derivatives is given the name "pterin." The generalized structure of pterins is represented as follows:

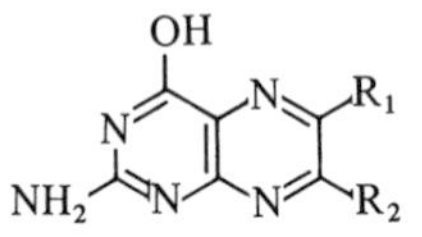

OH
N N R_1
NH_2 N N R_2

Pterin compounds were first identified in extracts of butterfly wings, but are now known to be widely distributed in animals, plants and bacteria (FORREST, 1962). The chemical methods involved in working with pteridines and determining structural formulas are beyond the scope of this work. The reader is referred to the reviews by ZIEGLER and HARMSEN (1969) and FORREST (1962) and to papers by FORREST and MITCHELL (1954a,b, 1955) and FORREST *et al.* (1959) for discussions of these methods.

Of primary interest for us is the occurrence of pterins as red pigments in the compound eye of *Drosophila*, and the presence of assorted other pterins in eye and other tissues, notably testes. The principal method of investigating pterins in normal and mutant *Drosophila* has been separation of the compounds by paper chromatography in alkaline solvent systems (typically ammonia-propanol), and detection by their fluorescent properties. This method, which is easily applicable to single flies or organs, was introduced by HADORN and MITCHELL (1951), and has been modified by various scientists (see GREGG and SCHMUCKER, 1965). There are certain problems in interpreting the work based on this method. For one thing, the various spots separated chromatographically cannot always be identified chemically, and there are even uncertainties regarding the structures of some of the well-studied ones. In addition, these compounds are labile, particularly in light, and the pterins identified by chromatography may be break-down products of the compounds actually present in living tissue. It is often possible only to assert that a mutant lacks a certain chromatographic spot or spots, or that it accumulates extra substances, without being certain of the chemistry or the actual metabolic lesion.

Pterins in Eye Color Mutants

HADORN'S and MITCHELL'S (1951) initial chromatographic investigation established that eye color mutants characterized by red eyes had a normal complement of pterins, but those mutants with an evident reduction in red pigmentation typically had altered pterin patterns. Thus *cinnabar (cn), cardinal (cd), vermilion (v),* and *scarlet (st)* had basically normal patterns, while mutants such as *brown (bw), white (w), sepia (se),* and *sepiaoid (sed)* have qualitatively altered patterns. An extensive series of other mutants have distinct quantitative differences in the pterin pattern. This group includes *purple (pr), prune* (pn^2), *garnet* (g^2), *light (lt)*, and a series of *white* alleles. In a more recent study along the same lines, GREGG and SCHMUCKER (1965) grouped the non-red eye color mutants into about a dozen classes based on pterin pattern. They were able to use these groupings to infer homologies between eye color mutants in different *Drosophila* species.

Some mutants are characterized by a general depression in level of all pterins. Among these, *bw* and *w* are the most extreme examples. One might conclude that these mutants are blocked in an early synthetic step leading to pterins. However, the newly emerged adults contain some non-pigment pterins (e.g., isoxanthopterin), which are lost shortly after eclosion, apparently in the meconium (HADORN and SCHWINCK, 1956a; BONSE, 1969). Other mutants, notably *se*, lack specific pterins but show a considerable accumulation of others. Thus, *se* forms none of the red pigments (collectively called drosopterins), but has a considerable accumulation of a yellow pigment, sepiapterin (FORREST and MITCHELL, 1954a, b). In the classical interpretation of genetic blocks, the accumulated compound is likely to be a precursor to the missing product. Analysis of the products that are missing and/or accumulated in a series of mutants can often provide strong evidence for the sequence of steps in a metabolic pathway. It might, therefore, be hoped that the variety of available eye color mutations in *Drosophila* would have facilitated analysis of the metabolic conversions of pterins and, in turn, led to the identification of the specific enzymatic defects associated with the various mutations. It has indeed been possible to postulate patterns of conversion among the pterins based on findings of this type (see HUBBY and THROCKMORTON, 1960; ZIEGLER and HARMSEN, 1969). However, crucial confirmation of conversions postulated on the basis of genetic defects requires demonstration, *in vitro*, of enzymes that catalyze the postulated reactions, and which are present in wild-type flies but absent or severely reduced in the respective mutants. Such confirmation is lacking in all but two instances, *rosy (ry)* and *maroon-like (ma-l)*, both involving the same enzyme.

Chromatographic analysis of these two mutants shows them to share the absence of a particular pterin spot identified as isoxanthopterin (HADORN and MITCHELL, 1951; HADORN and SCHWINCK, 1956a, b). FORREST *et al.* (1956) identified the one deficient enzyme in both cases as xanthine dehydrogenase (XDH) which, in addition to oxidation of xanthine and hypoxanthine, catalyzes the conversion of 2-amino-4-hydroxy pteridine to isoxanthopterin. Further details on these mutants are given in the chapter on XDH and related enzymes. Even in this case, the relationship of the enzyme to the metabolism of pterin pigments seems to be indirect, and is poorly understood. Thus the detailed pathways of pterin synthesis and interconversion in *Drosophila* remain unclear, and the enzymatic defects underlying the mutant eye color phenotypes is unknown in almost all cases.

The development of pigmentation in eye discs transplanted between various mutants was one of the most powerful tools in identifying some of the key gene products in the ommochrome pathway. Among many non-red eye color mutants tested, only *ry* and *ma-l* are non-autonomous in such transplant experiments (HADORN and SCHWINK, 1956a, b; HADORN and GRAF, 1958). Therefore, this approach has been of limited usefulness in elucidating pterin pathways. Implanted ry^2 eye discs develop normal pigmentation in wild-type hosts and in a variety of mutant hosts including *bw, ca, ma, ma-l, p,* rs^2, *se, w,* and w^a. The appearance of normal pigmentation in *w* and *bw* hosts supports the previously mentioned conclusion that these mutants are not blocked in an early step of pterin synthesis. Beyond this, the transplant experiments have not been particularly informative about the metabolic pathways leading to pterin pigments.

Other Mutations Affecting Pterins

Genetic variants that affect pterin metabolism without visibly altering eye color have also been reported by a number of others (HADORN and ZIEGLER-GÜNDER, 1958; HANDSCHIN, 1961; HADORN, 1958; HUBBY, 1962). Among the most interesting are mutants with tissue specific effects on the pterin complement. FABER and HADORN (1963) reported on sublines of *ma-l* with very different pterin distributions, apparently controlled by the genetic background and not by different *ma-l* alleles. One line, designated "w" had colorless testes with only traces of 2-amino-4-hydroxy pteridine, and no biopterin and sepiapterin. Their "g" subline had yellow testes that contained considerable quantities of 2-amino-4-hydroxy pteridine and sepiapterin and some biopterin. Drosopterin content in the eyes was similar for the two strains. HUBBY (1962) described a specific mutant designated *low isoxanthopterin (lix),* located on the X chromosome at 21.4±, which seems to affect only testes. This tissue contains very low concentrations of isoxanthopterin, and accumulates an unidentified blue fluorescing substance. Observations like this suggest that the pterin content of one tissue may vary more or less independently of that in other tissues of the same strain. This may well add to the confusion when attempts are made to interpret pterin patterns, particularly quantitative variations, obtained from whole fly extracts. The genes which control these tissue specific effects are unlikely to be structural genes for enzymes in the basic pathways of pterin synthesis and interconversion, but are of considerable interest as possible "regulators" of the expression of the structural genes. HADORN and ZIEGLER-GÜNDER (1958) report on the developmental pattern in a number of mutant lines. In addition, they noted a sex specific difference in that males contain much more isoxanthopterin than females. This appears to be due to accumulation in the testes, but this property is non-autonomous when testes are transplanted into a female host. Therefore, it would appear that the accumulation of isoxanthopterin by testes is, in some way, dependent on the surrounding environment.

In another kind of study, CHAUHAN and ROBERTSON (1966) and BARTHELMISS and ROBERTSON (1970) were able to isolate strains with qualitatively or quantitatively altered pterin contents, using the methods of population genetics. They showed that sublines with markedly divergent total pigment content and/or distinctive chromatographic patterns of pterins were readily obtained by repeated inbreeding starting from a large, genetically diverse population. One set of experiments involved pair matings of randomly selected sibs, repeated through several generations. A second experiment employed pair matings with the addition of artificial selection pressure. Several pair matings were set up each generation, and sibs from those cultures giving the most extreme divergence from normal pterin content (high or low) were used as parents for the next generation. Both procedures produced an assortment of biochemically distinct sublines. The response to selection suggests that the regulation of pterin pattern involves a considerable number of genes. The authors point out that the selection method they employed provides a convenient way to generate strains with defined biochemical differences. This might provide one alternative to the seemingly difficult task of determining the primary biochemical defect, starting with mutants having a visible phenotype.

Xanthine Dehydrogenase and Related Enzymes

Xanthine dehydrogenase (XDH), aldehyde oxidase and pyridoxal oxidase are discussed together in this chapter because they share certain features of genetic control, and they seem to be related in a number of ways. These three enzymes and at least six genetic loci, some with multiple alleles, are involved in one of the most complex and interesting gene-enzyme systems yet studied in any eukaryote. Reviews of the early work on this system have appeared elsewhere (GLASSMAN, 1965; GLASSMAN *et al.*, 1968). A more recent discussion may be found in URSPRUNG (1973).

In the section on pteridine based eye pigments and the genes affecting them, we mentioned that XDH is absent in two reddish-brown eye color mutants, *rosy* (*ry*; 3-52.0) and *maroon-like* (*ma-l*, 1-64.8). The early chromatographic work of HADORN and MITCHELL (1951) established that these two mutants lack a fluorescent spot identified as isoxanthopterin. Further work on these and an independently isolated mutant at the *rosy* locus (ry^2) was carried out by HADORN and SCHWINK (1956a, b). In addition to the absence of isoxanthopterin, these mutants are characterized by a slight accumulation of 2-amino-4-hydroxy pteridine, and lowered concentrations of the red pigments collectively known as drosopterins. It is the latter property that is presumably responsible for the dull reddish-brown eye color (orange in the presence of ommochrome mutants such as *vermilion* or *cinnabar*).

The specific enzyme defect shared by *ry* and *ma-l* flies was discovered by FORREST *et al.* (1956) during a series of studies designed to detect interconversions of pterins carried out *in vitro*. Various pterins were incubated with extracts of wild-type flies and assorted eye color mutants, and the mixtures were then chromatographed to detect and identify any products formed during the incubation. The only activity detected in this way was the conversion of 2-amino-4-hydroxypteridine into isoxanthopterin. Since it had previously been reported that XDH from other sources catalyzed this conversion, FORREST *et al.* (1956) investigated the possibility that *Drosophila* extracts contain a similar enzyme. They found their extracts could, in fact, oxidize xanthine to uric acid. They also confirmed that milk XDH catalyzes the conversion of 2-amino-4-hydroxypteridine to isoxanthopterin. Finally, they found that extracts of *ry* and *ma-l* flies did not catalyze either reaction. This is, of course, in good agreement with the initial observations that these mutants lack isoxanthopterin and accumulate 2-amino-4-hydroxypteridine. Thus, it is concluded that *ry* and *ma-l* mutants have in common the absence of XDH. A very satisfactory confirmation of this conclusion is supplied by the observation that treatment of wild-type larvae with 4-hydroxy pyrazolo (3, 4d) pyrimidine, an inhibitor of XDH, produces a good phenocopy of the *ry* eye color (KELLER and GLASSMAN, 1965; BONI *et al.*, 1967).

Before going into the details of the biochemistry and genetics of this system, it may be useful to review briefly the observations that first suggested a relationship between the three enzymes. Recognition of aldehyde oxidase and pyridoxal oxidase as related enzymes is essentially based on fortuitous discoveries that they too were deficient in *ma-l* (but not *ry*) mutants.

FORREST *et al.* (1961) carried out further biochemical characterization of the reactions catalyzed by extracts of wild-type and mutant flies, again using chromatography to identify the products of any reactions. In addition to some other differences, they noted that wild type and *ry* extracts oxidized pyridoxal (actually added as a potential cofactor) to pyridoxic acid, while extracts of *ma-l* flies did not catalyze this reaction. These results suggest the existence of a second, independent, enzyme, pyridoxal oxidase, affected by the *ma-l* locus. This was confirmed for a series of *ma-l* alleles by CHOVNICK *et al.* (1969).

Aldehyde oxidase was added to the list of enzymes controlled by *ma-l* by COURTRIGHT (1967). GLASSMAN (1965) had reported that XDH catalyzed the oxidation of benzaldehyde. However, when COURTRIGHT used benzaldehyde as a substrate in reaction mixtures intended to stain XDH in electrophoresis gels, a band with a mobility different from XDH was detected. The enzyme responsible for the new band was absent in *ma-l* extracts, but present in *ry* extracts. A variety of other aldehydes also serve as substrate for this enzyme. Thus, again, there is evidence for an independent enzyme, aldehyde oxidase affected by *ma-l* and not by *ry*. As we shall see below, the evidence for a close relationship between the enzymes is strengthened by the discovery of another locus, *low xanthine dehydrogenase* (*lxd*, 3-33) that affects all three enzymes (KELLER and GLASSMAN, 1964a).

Biochemistry

Xanthine Dehydrogenase

XDH catalyzes the oxidation of hypoxanthine to xanthine and thence to uric acid, and the oxidation of 2-amino-4-hydroxypteridine to isoxhanthopterin (FORREST *et al.*, 1956; GLASSMAN, 1965). Both reactions require NAD^+ as a cofactor (MORITA, 1958; GLASSMAN and MITCHELL, 1959; FORREST *et al.*, 1961; COLLINS *et al.*, 1971). XDH is conveniently assayed by following the change in fluorescence when 2-amino-4-hydroxypteridine is oxidized to isoxanthopterin (GLASSMAN, 1962a). This method is sufficiently sensitive to use on single flies. Less sensitive spectrophotometric assays are possible by following the reduction of NAD^+ at 340 nm (PARZON and FOX, 1964), or the production of uric acid. Extracts are usually treated with activated charcoal prior to assay, to remove inhibitory small molecules generally present in whole fly extracts (GLASSMAN and MITCHELL, 1959; PARZON and FOX, 1964). Detection in electrophoretic gels is possible with the tetrazolium reduction method used with other dehydrogenases (SMITH *et al.*, 1963; YEN and GLASSMAN, 1965).

Partial purifications of XDH have been reported by GLASSMAN and MITCHELL (1959) and PARZON and FOX (1964). In the former procedure, an 80000 $\times$g supernatant was treated with activated charcoal, fractionated by $(NH_4)_2SO_4$ precipitation, chromatographed on a $CaPO_4$ gel and precipitated again with $(NH_4)_2SO_4$. The purification achieved was 10–50 fold, and the yield was 50–75%. PARZON and FOX employed heating to 50° C, precipitation at pH 5.0, $(NH_4)_2SO_4$ cuts and chromatography on DEAE cellulose. The purification reported was about 500-fold, and the yield was 47%. However, COLLINS *et al.* (1971) were unable to obtain comparable purifications with this method. More recently, SEYBOLD (1973a, b) has reported a procedure employing $(NH_4)_2SO_4$ precipitation, heat treatment at 68° and chromatography on DEAE cellulose, hydroxyl-apatite and sephadex G-200. Purification of over 400-fold and a yield of about 10% were obtained, and the product appeared homogeneous by disc electrophoresis and electrofocusing.

Estimates of K_m's for the major substrates were reported by GLASSMAN and MITCHELL (1959). These are: 2-amino-hydroxy pteridine -6.7×10^{-6} M; hypoxanthine -2.5×10^{-5} M; and xanthine -2.1×10^{-5} M.

On the basis of hybrid enzyme formation in flies heterozygous for different electrophoretic variants, YEN and GLASSMAN (1965) conclude that the enzyme is at least a dimer. The molecular weight, estimated by a combination of gel filtration and velocity sedimentation in sucrose gradients, is about 250000 (GLASSMAN *et al.*, 1966). Electrophoresis in SDS acrylamide gels indicates subunits of a molecular weight near 150000 (SEYBOLD, 1973b).

GLASSMAN *et al.* (1968) have reported alternative forms of XDH not based on allelic differences. Although fresh fly extracts yield a single active XDH band on acrylamide gel electrophoresis, following several purification steps the enzyme elutes from DEAE-cellulose columns in two peaks. Each peak, when chromatographed again, elutes at the same NaCl concentration as before. Furthermore, the two forms isolated chromatographically can be distinguished electrophoretically. Strains which have genetically determined electrophoretic differences also yield two forms in the same relative positions, though shifted in absolute mobility. Only XDH-I is found in fresh crude extracts. Partial conversion of XDH-I to XDH-II appears to take place during purification. Once purified, XDH-I does not convert spontaneously to XDH-II, but conversion can be accomplished by incubating purified XDH-I with extracts of *ry* or *ma-l* flies at 30° for 1 hr. The conversion does not take place at 0°. The converting factor in *ry* extracts is heat labile, non-dialyzable and precipitable with $(NH_4)_2SO_4$.

XDH-I and XDH-II have similar K_m's for the principal substrates, similar (though not identical) sedimentation rates in sucrose gradients, and similar sensitivity to trypsin and heat inactivation. The exact mechanism of conversion is unclear. GLASSMAN *et al.* (1968) report that the changed chromatographic properties of XDH-II can be used to separate it from the proteins contaminating XDH-I, and thus achieve high specific activity preparations of the enzyme, but details are not given.

Pyridoxal Oxidase

Pyridoxyl oxidase catalyzes the conversion of pyridoxal to the corresponding acid, and does not require NAD^+ as a co-factor (FORREST *et al.*, 1961; COLLINS and GLASSMAN, 1969). The reaction can be assayed fluorometrically (COLLINS and GLASSMAN, 1969), or by coupling the oxidation of substrate to reduction of dichlorophenol indophenol (DCPIP) (DICKINSON, 1969). No electrophoretic stain has been reported, although coupling to reduction of tetrazolium might be possible. Work on single flies is difficult (COLLINS and GLASSMAN, 1969), but chromatographic separation of the product after incubation of substrate with fly extract, as in the method of FORREST *et al.* (1961) can be sensitive enough to use with single flies (DICKINSON, unpublished). Little purification or biochemical characterization of this enzyme has been done. The molecular weight has been estimated at about 250000 (GLASSMAN *et al.*, 1966).

Aldehyde Oxidase

Aldehyde oxidase catalyzes the oxidation of a number of aldehydes, including acetaldehyde, benzaldehyde and salicylaldehyde. It does not require NAD^+ as a cofactor (COURTRIGHT, 1967; DICKINSON, 1970; COLLINS *et al.*, 1971). It is conveniently assayed using acetaldehyde as a substrate, and coupling its oxidation to reduction of DCPIP (COURTRIGHT, 1967; DICKINSON, 1970). It is also detectable in electrophoretic gels by coupling to nitro blue tetrazolium reduction, using benzaldehyde as substrate (COURTRIGHT, 1967). The physiological substrate of the enzyme remains unknown. MADHAVEN *et al.* (1973) have shown that farnesol serves as a substrate, and have suggested a role in juvenile hormone metabolism. However, the normal development and viability of mutants lacking the enzyme indicates that such a role, if real, is not vital. The K_m for benzaldehyde is much lower than that for acetaldehyde (10^{-5} and 10^{-2} respectively), but the V max is greater for acetaldehyde (DICKINSON, 1969).

Purifications of aldehyde oxidase have been reported by COURTRIGHT (1967) and DICKINSON (1970). The latter procedure employs a precipitation at pH 5, batch chromatography on carboxymethyl-cellulose, precipitation with ammonium sulfate and successive chromatography on DEAE-cellulose and hydroxylapatite. It yields a product over 250-fold purified, with a recovery of about 15%. The product electrophoreses as a single band of protein on acrylamide gels. The purified enzyme has been used to produce an antibody that forms a single precipitin band when reacted against crude extracts (DICKINSON, 1970).

Like XDH, aldehyde oxidase appears to exist in two interconvertable forms. Electrophoresis shows a major and minor band that are under coordinate genetic control (COURTRIGHT, 1967). Although the conditions of conversion are not as well worked out as with the XDH forms, the relative intensities of the two bands can be changed by *in vitro* treatments such as aging or heat treatment at 50° C (DICKINSON, 1969 and unpublished). Similarity to XDH is also noted in that formation of hybrid enzyme in flies heterozygous for electrophoretic variants indicates that the enzyme is at least a dimer (DICKINSON, 1970), and the molecular weight is also about 250000 (COURTRIGHT, 1967).

Relationships of the Enzymes

The intimate genetic relationships to be discussed below make it important to establish that XDH, aldehyde oxidase and pyridoxal oxidase are distinct molecular species, and not just different activities of the same enzyme. A number of independent observations support this conclusion. XDH and aldehyde oxidase have distinct electrophoretic mobilities, and each can be altered by mutation independently of the other (COURTRIGHT, 1967; GLASSMAN *et al.*, 1968; COLLINS *et al.*, 1971). The three activities are separable from each other by chromatography on DEAE-cellulose (GLASSMAN *et al.*, 1968; COLLINS *et al.*, 1971). The antibody prepared against purified aldehyde oxidase does not affect the other two activities (DICKINSON, 1969, 1970). Finally as we shall see below, mutations are known which eliminate each of the enzymes without affecting the other two.

Genetics

The Rosy Locus

The *rosy* (*ry*, 3–52.0) locus appears to be a structural gene for XDH. Flies homozygous for *ry* contain no isoxanthopterin or uric acid, and they accumulate the precursors of these substances (HADORN and SCHWINK, 1956a, b; MITCHELL *et al.*, 1959). They contain no detectable XDH (FORREST *et al.*, 1956; GLASSMAN and MITCHELL, 1959), but aldehyde oxidase and pyridoxal oxidase are present in normal amounts (COURTRIGHT, 1967; COLLINS and GLASSMAN, 1969). The activity in ry^+ heterozygotes is about half the wild type level, and this proportional dosage response extends to stocks carrying three ry^+ genes (GLASSMAN *et al.*, 1962; GRELL, 1962a,b). GLASSMAN and MITCHELL (1959) showed that *ry* mutants not only lack XDH activity, but also do not contain detectable immunologically cross reacting material (CRM) related to XDH. YEN and GLASSMAN (1965) discovered several electrophoretic variants of XDH that map to the position of *ry* mutants. Furthermore, crosses of stocks with electrophoretic variants of XDH to *ry* always produce F_1 hybrids with only the enzyme form present in the non-*ry* parent. This amounts to failure of complementation and indicates that the electrophoretic variants are allelic to *ry*. Combined, these data make a strong case for *ry* as a structural gene for XDH.

The *ry* region has been the subject of one of the most detailed fine structure analyses done in *Drosophila*. SCHALET *et al.* (1964) used deletions in the *ry* region to screen for X-ray induced visible mutations and the lethals within a 0.5 map unit segment. Of 26 non-*ry* mutants recovered in this screen, 19 were tested and placed into seven complementation groups on both sides of *ry*. None of the non-*ry* mutations in the region had any effect on XDH. A large series of *ry* mutants was obtained by screening for the typical eye color when mutagenized third chromosomes were placed in combination with ry^2 or a deletion covering the region including *ry*. All newly isolated *ry* mutants that were lethal when made homozygous were found, by complementation tests, to extend into one or more of the flanking functional groups. All non-lethal *ry* mutants, whether isolated by screening over ry^2 or over deletions covering the whole region, fall into one complementation group. These observations indicate that there is only one cistron related to XDH in this region. A series of *ry* mutants that do not extend into flanking functional units were subjected to fine structure mapping by CHOVNICK *et al.* (1964; see also CHOVNICK, 1966). Chromosomes carrying allelic mutations at separate sites within a cistron can, when combined in heterozygotes, recombine between the mutant sites to produce a

(a) linked lethal system

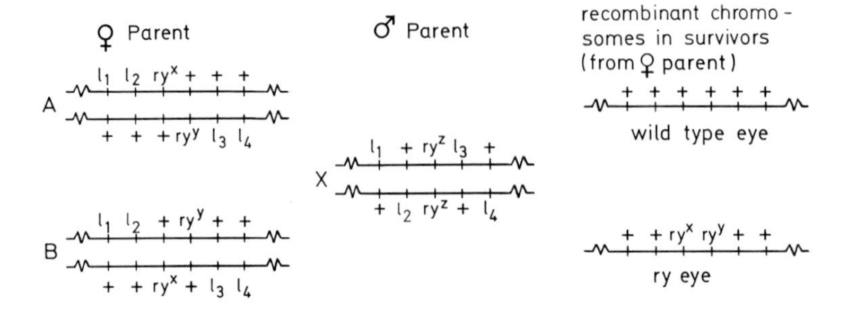

(b) purine selection system

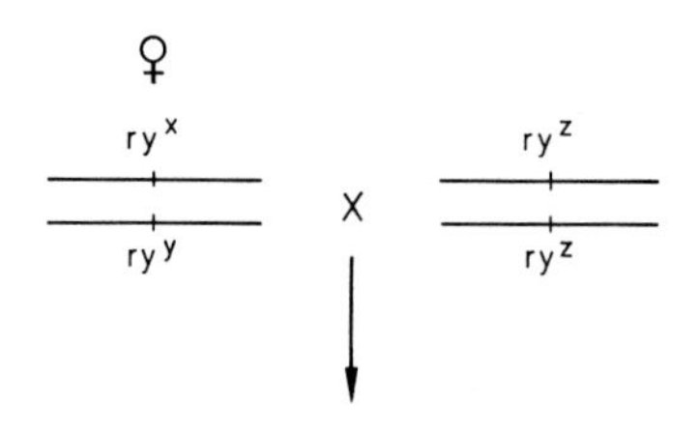

raise progeny on
medium with purine

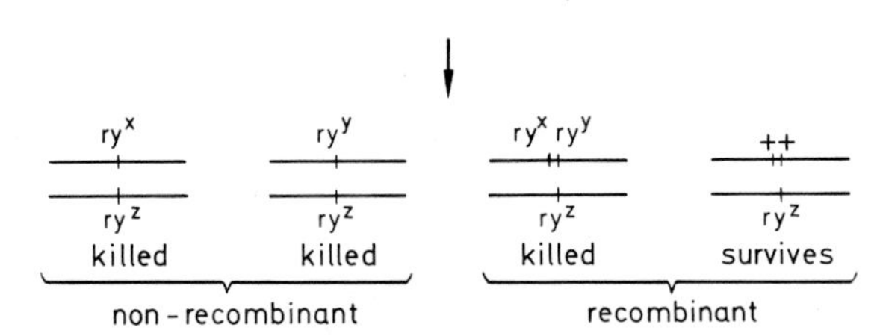

Fig. 12a and b. Two systems for fine structure mapping of the *ry* region. 12a (top) shows the linked lethal system of CHOVNICK *et al.* (1962, 1964). Each pair of *ry* alleles to be tested is built into females of types A and B. When crossed to the male shown, all progeny receiving non-recombinant chromosomes will be homozygous for one of the recessive lethals ($l_1 - l_4$). Recombinants that can survive are: $^1/_4$ of recombinants between l_1 and l_2; $^1/_4$ recombinants between l_3 and l_4; $^1/_2$ of recombinants between l_2 and l_3. Among the survivors recombinant between l_2 and l_3, there can be both double mutants and wild-types as shown at the right. Double mutants will be indistinguishable from single mutants, so only recombinations producing wild-type chromosomes are of interest. If the order within the *ry* cistron is as shown (ry^x to the left of ry^y), then wild-type recombinants will be found only among the progeny of type A females. If the order is reversed, these recombinants will be found among the progeny of type B females. Hence, the order of the *ry* alleles may be determined. From the frequency of wild-type progeny among the survivors and the known distances between l_1, l_2, l_3, and l_4, the map distance may be determined. In 12b (bottom), females heterozygous for two *ry* alleles are crossed to *ry* males. Purine is added to the medium on which the progeny are raised. Only wild-type progeny survive. The order of *ry* alleles may be determined by including outside markers. However, unlike the linked lethal method, this system detects wild-type chromosomes generated by processes that do not produce recombination of flanking markers (see text)

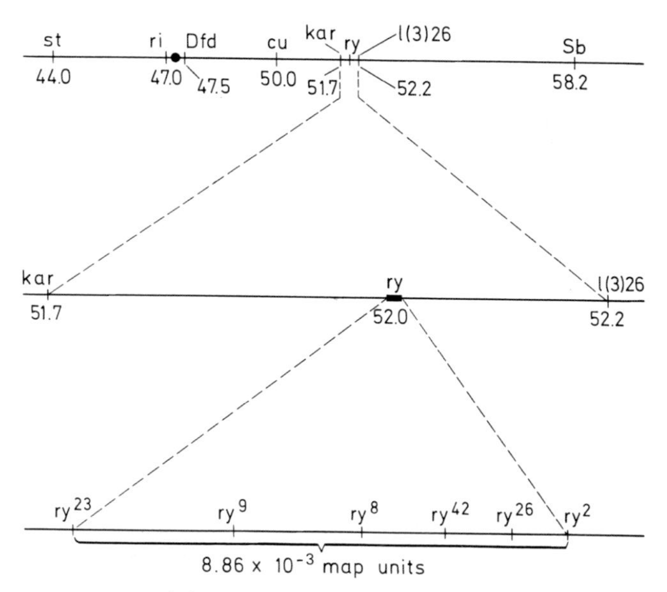

Fig. 13. Fine structure map of the *ry* region. At the top are shown the relative positions of several genes around *ry*. In the middle, the region between the closest flanking markers is expanded. The heavy bar represents the approximate extent of the *ry* cistron (the distance between the furthest separated alleles, 8.86×10^{-3} map units). At least seven other complementation groups are included within the region between *kar* and 1(3)26. The *ry* cistron is expanded at the bottom, showing at least six separable sites. Several other alleles have not been separated from ry^2. (Based on data of SCHALET *et al.*, 1964; CHOVNICK *et al.*, 1964)

wild-type allele. Thus, intracistronic recombinants can be detected as wild-type progeny in a mating of the type

$$♀\ ry^x/ry^y \times ♂\ ry^2/ry^2 .$$

Since all *ry* alleles belong to the same complementation group, wild-type progeny can arise only by recombination or by back mutation. Since wild-type progeny are never produced by females homozygous for single *ry* allele, back mutation is apparently not a significant factor.

Normally, a great number of progeny must be examined to detect recombinants on this fine level. However, CHOVNICK *et al.* (1962) devised a system of flanking recessive lethal markers such that only flies receiving a chromosome recombinant within a narrow region around *ry* survive (Fig. 12). Since most of the progeny die, resolution comparable to that obtainable by examining over 10^6 flies is possible. The arrangement of the flanking markers also reveals the order of the sites within the cistron. By this method, CHOVNICK *et al.* (1964) were able to show that 16 independent *ry* mutations were separable into at least six sites with a linear map (Fig. 13). This work was very important in establishing that functional units (cistrons) in higher organisms have a genetic fine structure comparable to that observed in bacteria and viruses.

The system for fine structure analysis used in this work is complex in that it depends on specially constructed stocks with the appropriate set of lethal markers linked to each allele to be tested. Furthermore, only those intra-cistronic events which are associated with recombination of flanking outside markers can be detected. The linked lethal method does have the advantage that it is in principle applicable to the analysis of any region of the genome, provided only that appropriate flanking lethals can be found or generated. A system which avoids the problems mentioned above is based on the fact that larvae lacking XDH are killed by purine in the medium (GLASSMAN, 1965; FINNERTY *et al.*, 1970). This system can be used to select for wild-type progeny generated by recombination between allelic mutations, including exceptional wild-type progeny that are not associated with recombination of outside markers. CHOVNICK *et al.* (1970, 1971) detected such progeny from flies heterozygous for different *ry* alleles (see also discussion of *ma-l* below). Their detailed analysis attributes the generation of a wild-type allele to gene-conversion events. These events resemble classical recombination events in that they occur only when two different alleles are present in combination, do not occur in *Drosophila* males and are suppressed by heterozygous rearrangements with breaks immediately flanking the mutant region. They differ in that no recombination of flanking markers is observed. BALLANTYNE and CHOVNICK (1971) have also shown that gene conversion events are not reciprocal. The demonstration of non-reciprocal events normally depends on tetrad analysis, i.e., recovery and testing of each of the four products of a single meiotic division. This is not possible in *Drosophila,* but use of compound chromosomes permits recovery of two of the four products, allowing a half tetrad analysis. This is the approach used by BALLANTYNE and CHOVNICK (1971). CHOVNICK *et al.* (1971) have proposed a model relating gene conversion to classical recombination, but a discussion is beyond the scope of this book.

The Aldox Locus

The structural gene for aldehyde oxidase (*Aldox*, 3-56.7) was identified and located in *D. melanogaster* using variants discovered by DICKINSON (1970). Electrophoretic variants and enzyme null mutants were independently mapped to the above position, and also behaved as alleles in complementation tests. As with XDH, flies heterozygous for two different electrophoretic alleles have a hybrid band of activity in addition to the two parental types. The null mutants have normal levels of XDH and pyridoxal oxidase. Heterozygotes for the null mutations have half the enzyme activity of homozygous normal flies. Using antibody produced against purified enzyme, DICKINSON (1970) showed that flies homozygous for the null mutant do not produce detectable CRM. In all of these respects, the *Aldox* and *ry* loci are completely analogous. COLLINS *et al.* (1971) report a mutation in which aldehyde oxidase is depressed but not absent, designated *low aldehyde oxidase (lao),* which maps in the same region but which they suggest is not allelic to the previously reported null mutant *($Aldox^n$).* Tests by DICKINSON (unpublished) suggest that this mutation does not complement $Aldox^n$, and is in fact allelic. COURTRIGHT (1967) reported electrophoretic variants of aldehyde oxidase in *D. simulans* and mapped the controlling gene to 3-74.5. IMBERSKI (1971) found that in *D. hydei,* some strains have a three-banded pattern resembling that seen in *D. melano-*

gaster individuals heterozygous for different electrophoretic alleles. However, all flies in these *D. hydei* strains have three bands, and individuals with only one band are never segregated. This suggests that the *Aldox* gene is duplicated in *D. hydei.*

The Low Pyridoxal Oxidase Locus

COLLINS and GLASSMAN (1969) identified a locus thought to be a structural gene for pyridoxal oxidase, and designated it *low pyridoxal oxidase (lpo,* 3-57 ±). The mutant was found in a survey of stocks for variations in pyridoxal oxidase activity. Homozygous mutant flies contain less than 5% of normal pyridoxal oxidase activity, but XDH and aldehyde oxidase are normal. Heterozygotes show a proportional response, and this is taken as evidence that this is a structural locus for the enzyme. However, no electrophoretic variants are known, and no immunological work has been done, so the evidence is less complete than for *ry* and *Aldox*. The map position is intriguingly close to that of *Aldox,* but fine structure analysis of this region has yet to be done.

The Maroon-Like Locus

The *maroon-like* (*ma-l,* 1-64.8) locus is characterized by mutants that resemble *ry* mutants in eye color, chromatographic pattern of pteridines, and absence of XDH (FORREST *et al.*, 1956; GLASSMAN and MITCHELL, 1959). We have already mentioned evidence which distinguishes *ma-l* from *ry* in that the former also lacks pyridoxal oxidase (FORREST *et al.*, 1961) and aldehyde oxidase (COURTRIGHT, 1967). The conclusion that *ma-l* lacks at least three separate enzymes has been extensively supported by subsequent work (see DICKINSON, 1970; COLLINS *et al.*, 1971; CHOVNICK *et al.*, 1969).

There are other important ways in which *ma-l* differs from *ry*. The proportional dosage response characteristic of *ry* and *Aldox* fails in *ma-l,* in that heterozygotes contain normal levels of both XDH and aldehyde oxidase (GLASSMAN *et al.*, 1962; DICKINSON, 1970). Furthermore, *ma-l* mutants contain CRM related to both enzymes (GLASSMAN and MITCHELL, 1959; DICKINSON, 1970). Thus *ma-l* mutants seem to make the primary products of the *ry* and *Aldox* loci, but they are enzymatically inactive.

There is an interesting maternal effect that influences expression of the *ma-l* phenotype. Genetically *ma-l* progeny of *ma-l*/+ females have wild-type eye color and a low but detectable level of XDH (GLASSMAN and MITCHELL, 1959) (Fig. 14). The maternal effect disappears with aging of the cultures, but this appears to be related to changes in the medium rather than the age of the mothers. The *ma-l*/+ females are said to transmit a $ma\text{-}l^{+}$ substance to their progeny, but it is not yet clear what this substance is. We will further discuss this below in conjunction with experiments on *in vitro* complementation between the products of various loci.

Detailed analysis of *ma-l* has shown this locus to be more complex than *ry*. GLASSMAN and PINKERTON (1960) reported that a second mutation, $ma\text{-}l^{bz}$ has a similar eye color, lacks XDH, is maternally affected and could not be separated from *ma-l* by recombination among 5000 progeny. However, $ma\text{-}l/ma\text{-}l^{bz}$ heterozy-

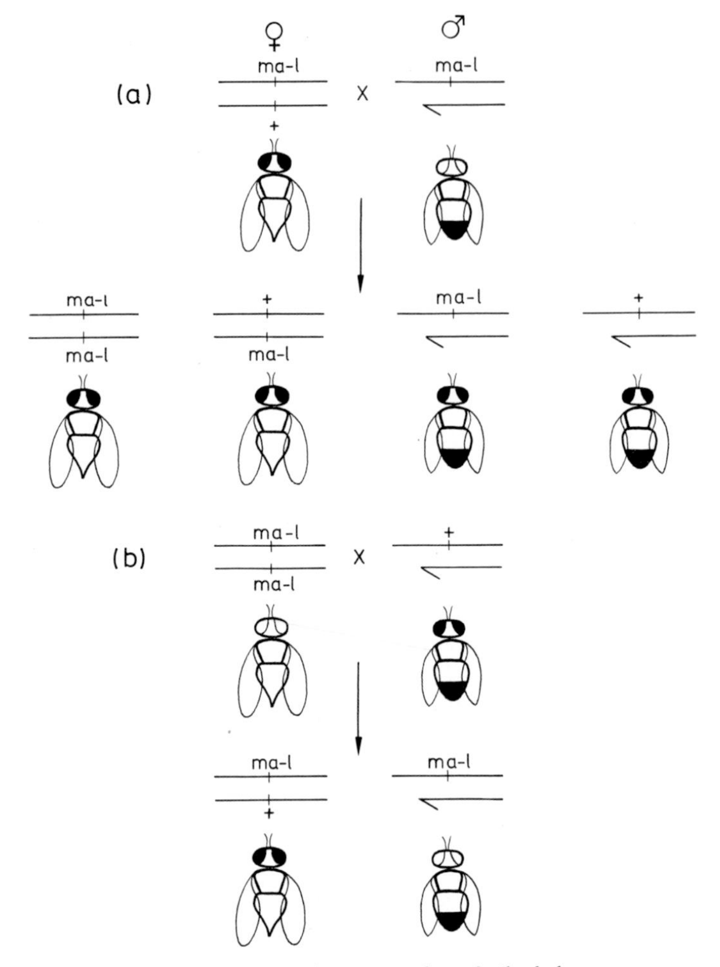

Fig. 14a and b. The maternal effect on inheritance of *ma-l*. Shaded eyes represent wild-type, unshaded are *ma-l*. The X-chromosome is represented as a straight rod, and the Y is shorter and has a hook. In cross (a), all progeny have wild-type eyes, even though half the females are homozygous for the mutation and half of the males are hemizygous. The genotype is expressed in the normal way if the female parent was homozygous for the mutation (cross b)

gotes have normal eye color and about 5% of the wild-type level of XDH. URSPRUNG (1961) confirmed this complementation and reported that the concentration of drosopterins approaches the wild-type level.

CHOVNICK *et al.* (1969) examined 37 *ma-l* mutations, both X-ray and chemically induced. Of these 19 were fully viable while 18 were homozygous or hemizygous lethal. As with *ry*, all of the *ma-l* lethals could be shown to be deletions extending into adjacent loci whose mutants are lethals. The 19 viable mutations fell into five complementation groups (Fig. 15) Group I mutants do not complement with any of the other groups. Group II complements group V, but not III and IV. Groups III and IV complement each other, and Group V, but do not complement Groups I and

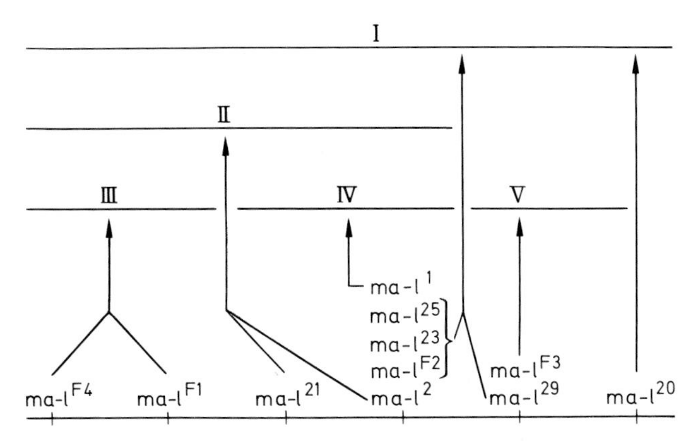

Fig. 15. Complementation map of the *ma-l* cistron compared to the fine structure map (CHOVNICK *et al.*, 1969). The Roman numerals identify bars representing the extents of the five complementation groups. Groups whose bars overlap do not complement. The arrows indicate assignments of various alleles to the complementation groups, and the relative positions of these alleles along the linear fine structure map are indicated at the bottom

II. This complex pattern of complementation is explainable by at least three models (CHOVNICK *et al.*, 1969). 1. There may be several linked cistrons involved in the production of the *ma-l*$^{+}$ phenotype. In this model, group III, IV, and V mutants would represent mutations in the various cistrons, while group I and II mutations would be deletions extending into more than one cistron. 2. The second model also assumes adjacent cistrons, but further assumes that they are transcribed as a unit (as an operon), and that group I and II mutations include polar mutations as well as deficiencies. 3. The third model propose a single cistron exhibiting allele complementation, presumably by some interaction at the product level. A genetic fine structure analysis (see below) shows that at least some goup I and II mutants are site mutants, not deletions, and hence model 1 is eliminated. The order of site mutants in complementation groups III, IV, and V is consistent with the adjacent cistron model, and the order of group I and II mutants is also consistent with the polar nature of their complementation, but this data does not serve to distinguish between models 2 and 3.

CHOVNICK *et al.* (1969) also examined the XDH activity in flies heterozygous for various *ma-l* alleles. Those heterozygotes which show no complementation on the basis of eye color also have no XDH. Combinations of *ma-l* alleles that complement for eye color produce XDH ranging from 5% to 25% of wild-type. In contrast, even the allele combinations that have considerable XDH activity do not have detectable pyridoxal oxidase activity. This last observation is more consistent with the view of *ma-l* as a single cistron, and favors the allele complementation model (model 3).

Additional evidence has been obtained favoring the single cistron allele complementation model of complementation at the *ma-l* locus. CHOVNICK *et al.* (1969) reason as follows: If allele complementation occurs in this system as in others, it involves formation of dimers or higher aggregates containing subunits of both mutant types. If mutants *ma-l*^{x} and *ma-l*^{y} complement each other and *ma-l*^{z} complements neither, but does make some product (as evidenced by complementation to other alleles), then the product of a *ma-l*^{z} allele placed in combination with both *ma-l*^{x} and *ma-l*^{y} might compete for aggregation with the complementing pair, and decrease the amount of active aggregate. Thus flies of the genotype *ma-l*^{x}/*ma-l*^{y}/*ma-l*^{z} would make less active product (and hence less XDH) than simple *ma-l*^{x}/*ma-l*^{y} heterozygotes. The polar mutation model makes no such prediction. The test was conducted by synthesizing an attached-X chromosome carrying *ma-l*^{F1} and *ma-l*^{1} (corresponding to *ma-1*^{x} and *ma-l*^{y} in that they complement) and a duplication on a Y chromosome carrying *ma-l*^{2} (corresponding to *ma-l*^{z} in that it does complement group V mutants). Attached-X females ($\overline{XX}$/Y) carrying these three alleles served as the *ma-l*^{x}/*ma-l*^{y}/*ma-l*^{z} flies desired for the test, while attached-X females with a Y chromosome carrying a similar duplication from the X chromosome, but with the *ma-l* locus deleted, served as the *ma-l*^{x}/*ma-l*^{y} heterozygotes. Careful assays of the XDH levels in these flies shows that the extra *ma-l* allele does indeed depress activity in about the expected ratio. Thus, the single cistron, allele complementation model is favored.

FINNERTY and CHOVNICK (1970) obtained independent evidence for the allele complementation model. They performed large scale recombination experiments from which double mutants carrying *ma-l*^{F4} (group III) and *ma-l*^{F3} (group V) were recovered. By the three cistron model, this double mutant should still complement group IV mutants. The allele complementation model does not require this predication and, indeed, one might expect the product of a gene mutant at two sites to be so altered that it could not complement. The appropriate tests indicate that the double mutant does not complement *ma-l*^{1} (group IV), and hence again supports the single cistron model.

CHOVNICK'S group (FINNERTY *et al.*, 1970; SMITH *et al.*, 1970; CHOVNICK *et al.*, 1971) has also carried out studies on intra-cistronic recombination and gene conversion events in the *ma-l* cistron. The purine selection system described in the section on the *ry* locus was also used here. Females heterozygous for a pair of *ma-l* alleles were mated to tester males, and the progeny were raised on purine enriched medium, so that only *ma-l*$^+$ progeny survived, (FINNERTY *et al.*, 1970). An estimate of the total progeny screened in this way is obtained by omitting purine from some cultures and counting the total progeny they yield. The relative position of a pair of *ma-l* mutants can be determined from the pattern of recombination of outside markers associated with *ma-l*$^+$ progeny. For details, consult FINNERTY *et al.* (1970).

A total of 61 *ma-l*$^+$ progeny were recovered from a screen that effectively examined 6×10^7 zygotes. Of these, 25 (41%) were associated with recombination of outside markers, and are assumed to represent *ma-l*$^+$ chromosomes generated by recombination between *ma-l* site mutants. The map produced by analyzing the association of *ma-l* recombinants with outside markers is linear with at least six separable sites (Fig. 15). As mentioned previously, the recombination map is co-linear with the complementation map.

Thirty six of 61 *ma-l*$^+$ progeny in the screen were not associated with recombination of flanking markers, and were designated *ma-l*$^+$ exceptionals. Various crosses used to preserve and/or test the *ma-l*$^+$ chromosomes eliminate the possibility that dominant or recessive autosomal mutations or recessive X-linked mutations acting as suppressors can account for the *ma-l*$^+$ phenotype. Closely-linked suppressors or second-site mutations within *ma-l* are not eliminated by these crosses, but in a random sample of *ma-l*$^+$ exceptionals, all showed full wild-type levels of XDH, which would be very surprising for suppressors or second-site mutations. A number of considerations also argue against double recombination or spontaneous back mutation, as the origin of the *ma-l*$^+$ exceptionals (FINNERTY *et al.*, 1970) and their origin as gene conversion (i.e., non-reciprocal) events is favored.

Use of attached-X chromosomes permits examination of two of the four products from a meiotic event, and represents a compromise between the full tetrad analysis, which gives all four products, and analysis in other systems, which gives only one (SMITH *et al.*, 1970). Using this advantage of the attached-X system, SMITH *et al.* (1970) obtained strong evidence that all *ma-l*$^+$ exceptionals arise by non-reciprocal conversion events.

The Low Xanthine Dehydrogenase Locus

Another locus affecting XDH, *low xanthine dehydrogenase* (*lxd*, 3-33), was discovered by KELLER and GLASSMAN (1964a) in a survey in which they compared XDH activity levels in a series of *D. melanogaster* strains. A similar and perhaps identical mutation at the same locus was recovered by continued selection for low XDH activity through several generations of pair matings, starting from a genetically heterogeneous population (KELLER and GLASSMAN, 1964c). The mutant has 20–25% of normal XDH activity. It also proves to have even more severely depressed levels of aldehyde oxidase (COURTRIGHT, 1967) and pyridoxal oxidase (GLASSMAN, 1965). In both cases, the electrophoretic mobility and heat stability characteristics indicate that the small amount of enzyme present is normal. Like *ma-l*, *lxd* appears to produce XDH CRM in excess of the amount of active enzyme, and comparable to the amount found (as active enzyme) in wild-type flies (GLASSMAN *et al.*, 1968). Also like *ma-l*, *lxd* heterozygotes have the full wild type level of XDH activity. However, COURTRIGHT (1967) found that the *lxd*/+ heterozygotes contain about 60% of wild-type aldehyde oxidase activity, and he could find no aldehyde oxidase CRM in *lxd*. This last conclusion is questionable in view of his failure to detect aldehyde oxidase CRM in *ma-l*, (subsequently demonstrated by DICKINSON, 1970), and has not been confirmed in recent unpublished experiments (WEISS, 1972).

KELLER and GLASSMAN (1964a) placed *lxd* on the third chromosome, using dominant markers. It was further mapped by testing progeny of recombinant males obtained from the cross ♀ + + + *lxd* +/*ru jv se* +*by* x ♂ *ru jv se* + *by* and backcrossed to + + + *lxd* + females. The resultant map position is 3-33±.

The Cinnamon Locus

Recently, BAKER (1973) has reported a mutation at a new locus, *cinnamon* (*cin*, 1-1±), which causes a deficiency of XDH and has a variety of other effects. Homozygous *cin* females are almost completely sterile when mated to *cin* males, but are fertile with *cin*$^+$ males. Heterozygous *cin*/*cin*$^+$ females are fertile with either kind of male. It appears that *cin*$^+$ performs a function vital to survival of the embryo, but that this function can be fulfilled either by action of the maternal genome (*cin/cin* progeny of *cin*$^+$/*cin* females survive), or the embryos own genome (*cin*$^+$/*cin* progeny of *cin/cin* females survive). Occasional survivors are found among homozygous progeny of homozygous females. These rare individuals have an eye color resembling that found in *ry* and *ma-l* flies, and chromatographic analysis reveals a qualitatively similar alteration of pterin pattern. Homozygous progeny of heterozygous females have the normal eye color. Thus, this locus displays a maternal effect similar to that found with *ma-l*, but affecting both eye color and viability. XDH is absent in *cin/cin* flies, but pyridoxal oxidase and aldehyde oxidase apparently have not been examined. The lethal effect of *cin* is almost certainly not due to the absence of XDH, since *ry* and *ma-l* have no such effect. Thus *cin*, like *ma-l* and *lxd*, must affect some factor(s) in addition to XDH.

Relationships between the Loci

The fact that two loci, *ma-l* and *lxd*, affect three different enzymes raises interesting questions about their functions. There are at least three models that might apply to one or both of these loci (see GLASSMAN, 1965; GLASSMAN *et al.*, 1968). The first possibility is that *ma-l* and/or *lxd* are regulatory loci that control the transcription (or translation) of the structural genes for the enzymes. This model is not favored by the fact that mutants at these loci produce CRM related to both XDH and aldehyde oxidase, and hence the structural genes are apparently active but the products fail to become enzymatically active. The alternative models suggest that these loci are involved in the production of either a cofactor needed in common by the three enzymes, or protein subunits shared by the three enzymes.

What seemed to be a handle for investigating the contribution of *ma-l*$^+$ was provided when GLASSMAN (1962b, c) discovered that extracts of *ry* and *ma-l* flies mixed together and incubated for a time at 37° produce small amounts of active XDH. Neither extract incubated alone does so. This *in vitro* complementation defines a *ry*$^+$ substance (present in *ma-l*), and a *ma-l*$^+$ substance (present in *ry*), and permits one to investigate the properties of each by using the complementing activity as an assay. The *ry*$^+$ substance is assumed to be the inactive product of the XDH structural gene, and hence is probably identical to the CRM detected in *ma-l* flies. The *ry*$^+$ substance is heat-labile and trypsin-sensitive (GLASSMAN, 1966). It has a molecular weight of about 250000, as do active XDH and CRM (GLASSMAN *et al.*, 1966). From this observation, it appears unlikely that the *ry*$^+$ substance lacks any substantial subunit normally supplied by *ma-l*$^+$. It is also reported that the double mutant *ma-l, lxd* still contains the *ry*$^+$ factor.

Assuming *ry*$^+$ factor is identical to CRM, the *ma-l*$^+$ substance is possibly of greater interest. Surprisingly, it too appears to be a high molecular weight protein, although it is more heat-stable and trypsin-resistant than *ry*$^+$ factor. The molecular weight is again about 250000 (GLASSMAN *et al.*, 1966). This precludes any simple addition of *ry*$^+$ and *ma-l*$^+$ substance during *in vitro* complementation, but leaves exchange of subunits or co-factors or some kind of catalytic modification as viable alternatives. GLASSMAN (1965) believes the latter is ruled out by the kinetics of complementation, which is bimolecular. One distinct possibility is that the *ma-l*$^+$ factor detected by *in vitro* complementation is another enzyme (e.g., pyridoxal oxidase or aldehyde oxidase) which can exchange subunits or co-factor with the *ry*$^+$ factor. This interpretation is consistent with the finding that *lxd, ry* flies do not contain *ma-l*$^+$ complementing factor (KELLER and GLASSMAN, 1964a) and, of course, are strongly reduced in both of the above enzymes. COURTRIGHT (1967) pointed out that aldehyde oxidase closely resembles *ma-l*$^+$ complementing factor in size, heat-stability, trypsin-resistance and genetic regulation. Complementation tests with *ry aldox*^{n}, *ry lpo* and *ry aldox*^{n} *lpo* as potential *ma-l*$^+$ donors should provide a test of this possibility. If one or both enzymes are the *ma-l*$^+$ substance, then the strains lacking those enzymes (in addition to XDH) should fail to complement *ma-l (ry*$^+$*)* extracts.

The maternal effect in the expression of the *ma-l* phenotype might also provide some insight into the nature of the *ma-l* substance (note, however, that the product of the *ma-l*$^+$ gene, the *ma-l*$^+$ complementing factor and the *ma-l*$^+$ maternal factor are not necessarily the same thing, or at least not in the same form). The maternal effect can be explained if *ma-l*$^+$ substance (in some form) is stored in the egg, or if XDH itself is stored. Either mechanism could provide the progeny with the small amount of XDH activity needed to produce the wild-type eye color. SAYLES *et al.* (1973) have detected *ma-l*$^+$ *in vitro* complementing factor in eggs, but this still does not establish the identity of the two *ma-l*$^+$ factors. XDH itself seems to be ruled out as the maternal substance because it is present in low amounts in the egg (GLASSMAN and McLEAN, 1962). A more convincing test is obtained genetically (Fig. 16). When attached-X ($\overline{XX}$/Y) females are mated to *ma-l* males, all male progeny are genetically *ma-l*. But all are maternally affected if the females are *ma-l*$^+$. When the attached-X females are *ry/ry* (and hence lack XDH), the male progeny are still maternally affected, so XDH cannot be the maternal factor.

Aldehyde oxidase, on the other hand, is accumulated in eggs (COURTRIGHT, 1967; DICKINSON, 1971). However, male progeny of $\overline{XX}$/Y, *aldox*^{n}*/aldox*^{n} females are still maternally affected (DICKINSON, 1970), so this enzyme cannot be the maternal substance either. This does not rule out aldehyde oxidase as the *ma-l*$^+$ *in vitro* complementing factor. Indeed, KELLER and GLASSMAN (cited in GLASSMAN, 1965) report that $\overline{XX}$/Y, *lxd/lxd* females also produce maternally affected male progeny, even though it is known that *lxd* flies do not contain the *ma-l*$^+$ *in vitro* complementing factor.

The linkage relationships between the loci are interesting. The two loci that affect all three enzymes are distant from each other and from the respective structural genes. The structural genes, however, are all within a 5-unit segment of the 3rd chromosome. DICKINSON (1970) has shown by direct recombination that *ry* and *Aldox* are indeed separated by about the distance indicated from independently obtained map positions. The reported positions of *Aldox* and *lpo* are much closer. However, the mapping precision, particularly for *lpo*, is not great. A direct measurement of the separation of these two loci would be interesting. The distance of *ry* from the other two loci, and the existence of apparently unrelated functions much closer to *ry* (SCHALET *et al.*, 1964) do not support an operon-like model for this system. One attractive possibility is that the three structural genes are related in an evolutionary sense through gene duplication. This could explain the similar

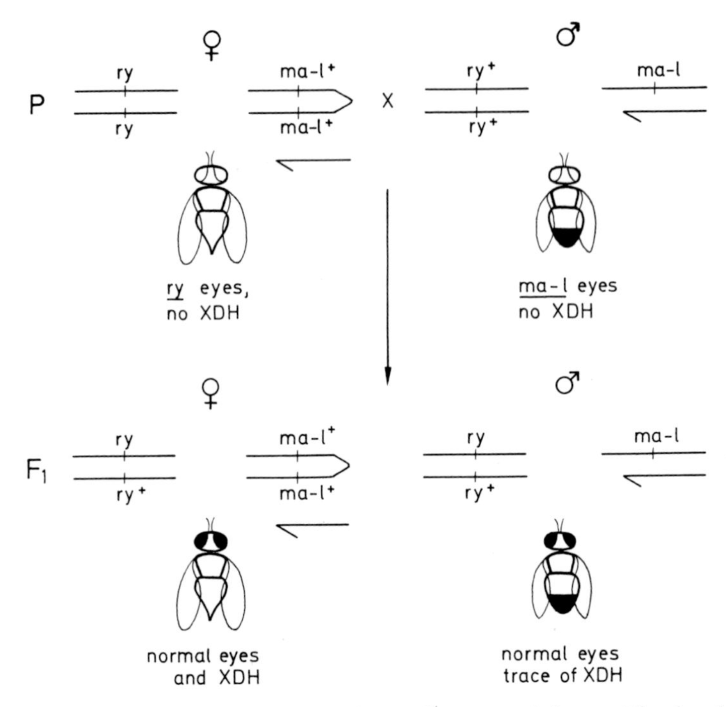

Fig. 16. Demonstration that XDH is not the *ma-l*$^+$ maternal factor. The female parent is homozygous for *ry* and hence has no XDH. Nevertheless, the F 1 males have wild-type eyes and a trace of XDH, even though they are hemizygous for *ma-l*. The use of attached X females (XX/Y) results in F 1 males having received their X-chromosome from their father. The same test may be performed with the female parent homozygous for *Aldox*^{n} with the same result, thus eliminating aldehyde oxidase as a possible *ma-l*$^+$ maternal factor

sizes, common elements of genetic control, and similar (even overlapping) activities (COLLINS *et al.*, 1971). Clarification of this relationship, and of the roles *ma-l* and *lxd*, may await purification of sufficient quantities of the enzymes to permit direct comparisons of amino acid content or sequence. ZOUROS and KRIMKAS (1973) report that the genes for XDH and aldehyde oxidase are also linked in *D. subobscura*. They also found an apparent linkage disequilibrium (i.e., one XDH allele is non-randomly associated with certain aldehyde oxidase alleles).

Development and Regulation

Developmental changes in enzyme activity and tissue localizations have been studied to some extent for both XDH and aldehyde oxidase, although no such information is available for pyridoxal oxidase. In view of the other similarities between these enzymes, it would be most interesting to know whether they are controlled coordinately during development. The information available does not

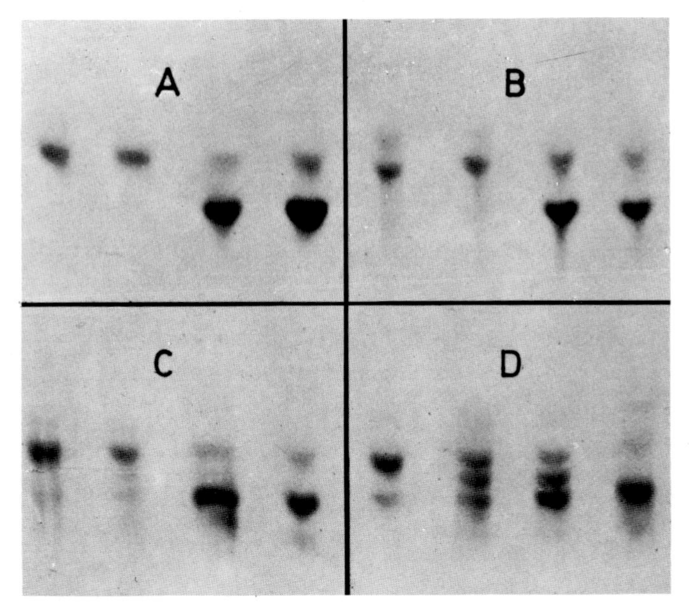

Fig. 17 A–D. The appearance of a hybrid aldehyde oxidase as an indicator of activity of the paternal gene. Extracts of staged larvae were arranged and electrophoresed in groups of four. The parental types are on the extreme left and right of each group, and reciprocal hybrids are between, each adjacent to the parental type of the maternal parent. The ages of the groups were: (A) eggs, 0–12 hrs; (B) eggs and first instar, 12–24 hrs; (C) first instar, 24–36 hrs; (D) first and second instars, 36–48 hrs. The hybrid band indicating production of paternal gene product is clearly visible in group D and barely detectable in group C. This indicates synthesis beginning at about 24 hrs, near the time of hatching

favor this interpretation, but is probably inadequate to rule out the possibility rigorously.

The developmental history of aldehyde oxidase has been worked out by DICKINSON (1971). As mentioned above, a considerable quantity of enzyme is found stored in the egg. There is no net increase until about the time of hatching. Furthermore, embryos that are heterozygous for two alleles determining electrophoretically different forms of the enzyme contain only the maternally inherited form up to about this same time (Fig. 17). Likewise, the progeny of a mating between heterozygous females and males homozygous for one of the electrophoretic forms do not show any dilution of the hybrid enzyme pattern until after hatching. All of this evidence supports the conclusion that no new synthesis of active aldehyde oxidase occurs until about the time of hatching. During larval development, there is a rapid net increase in total activity, but it is outstripped by the growth of the larvae, so the specific activity declines. The specific activity rises again at the time of pupation. There is a plateau or slight dip through much of the pupal period, and then another rapid increase in activity leading up to eclosion.

Dissection of third instar larvae indicates that the bulk of the enzyme is in the hypodermis, gut, and Malpighian tubules. Muscle contains no activity and fat body has relatively weak activity. Details of the distribution can be visualized using a

Fig. 18. Histochemical localization of aldehyde oxidase in larvae. The dark stain indicates areas of high enzyme activity. Most of the larval gut and associated structures are included. The anterior end is at the top, with salivary glands extending to each side. Malpighian tubules branch off near the bottom

specific histochemical stain basically like that used on electrophoretic gels (DICKINSON, 1971). The pattern in the gut is particularly interesting in that it shows the distribution of aldehyde oxidase to be highly differentiated (Fig. 18). For example, the cells of the ventriculous and proximal portions of the gastric cecae are heavily stained, while morphologically similar cells in the distal portion of the cecae have absolutely no activity. Similarly, the hindgut has a ring of strong activity just posterior to the insertion of the Malpighian tubules, and another near the extreme posterior end, but no activity in the intervening region.

In adults, aldehyde oxidase activity is concentrated in gut, Malpighian tubules and reproductive organs. Females contain 150–200% of the male activity, much of the difference being due to a large accumulation in ovaries (consistent with the high activity stored in eggs). As in larvae, the distribution is highly differentiated and reproducible (Fig. 19).

The specificity of the tissue distribution in both larvae and adults makes this a favorable system for investigating the regulation of differential gene expression. DICKINSON (1971, 1972, 1974) has obtained evidence of genetically determined alterations of the standard pattern, both with respect to distribution and timing. This suggests that it will be possible to identify and study genes concerned with the specific developmental expression of this enzyme. This could provide a badly needed genetic handle for the study of gene regulation in the context of eukaryotic development.

GLASSMAN and MCLEAN (1962) have reported some developmental information concerning XDH. They find little or no XDH in eggs, a steady increase in larval stages and a U-shaped-curve through the pupal stages. SAYLES *et al.* (1973) detected at least some accumulation of XDH by about 4 hours after fertilization. Enzyme appeared at the same time in *ry*/+ progeny of *ry*/*ry* females, ruling out any contribution of stored, inactive enzyme or maternal messenger. This is one of the earliest detected gene activations known in *Drosophila*. This is rather different from the aldehyde oxidase curve described above, but is perhaps compatible with coordinate synthesis of the two enzymes if XDH is less stable and declines during periods of low synthesis, and aldehyde

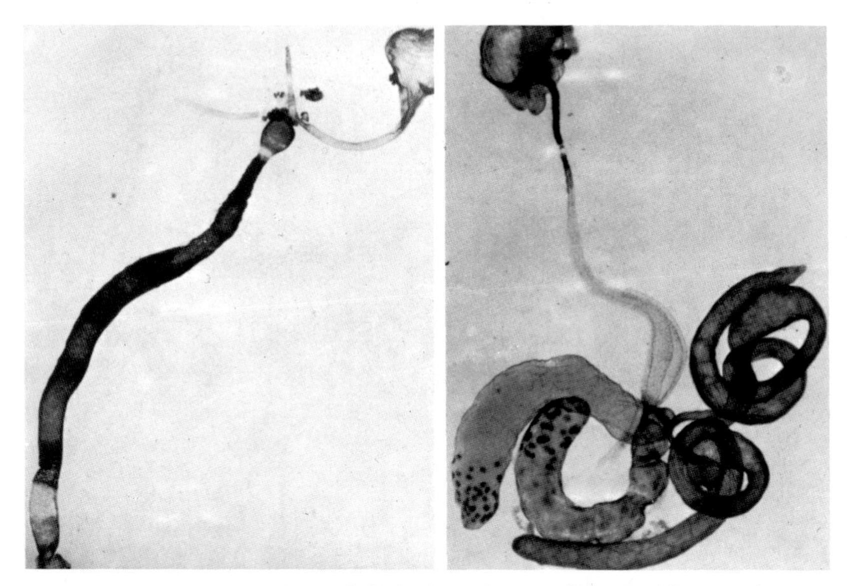

Fig. 19. Examples of tissue specificity of aldehyde oxidase in adults. Darkly stained areas contain high enzyme activity. The anterior midgut is at the left. At the right are male internal genetalia

oxidase is more stable and can be retained. Thus XDH activity is low in eggs and declines in the first half of the pupal period, while aldehyde oxidase activity is steady but not increasing during the same periods. A little early synthesis of aldehyde oxidase might go undetected in the presence of the large amount of stored enzyme.

URSPRUNG and HADORN (1961) have investigated the tissue distribution of XDH by assaying dissected organs. The relative activities, given in arbitrary units are: testes-0, Malpighian tubules-15, gut-25, fat body-44, hemolymph-44 and remainder-90. However, COLLINS *et al.* (1970) find almost half of the total activity in fat body, with Malpighian tubule being the next largest source of the enzyme. While the dissection experiments employed in these two papers do not provide the resolution available with the histochemical work on aldehyde oxidase, the gross distribution of the two enzymes does not seem to be parallel. This is most notable in the case of fat body which contains a large proportion of total XDH but a very small proportion of aldehyde oxidase. Again, this does not provide definitive proof that the two are not coordinately controlled. It must first be established that the enzyme present in a given tissue was synthesized there.

Autonomous synthesis has been fairly well established in the case of aldehyde oxidase. DICKINSON (1971) showed autonomy in ovary transplants between normal and *Aldox*^{n} flies (Fig. 20). JANNING (1972, 1974a, 1974b) used an unstable ring X chromosome to produce gynandromorphs that were mosaic for XX and XO tissue. The normal rod X used in these flies carried the *ma-l* mutation. Patches of tissue derived from cells that lost the ring X (*ma-l*$^+$) were therefore hemizygous for the mutation. Using the histochemical stain for aldehyde oxidase, he found patches of positive and

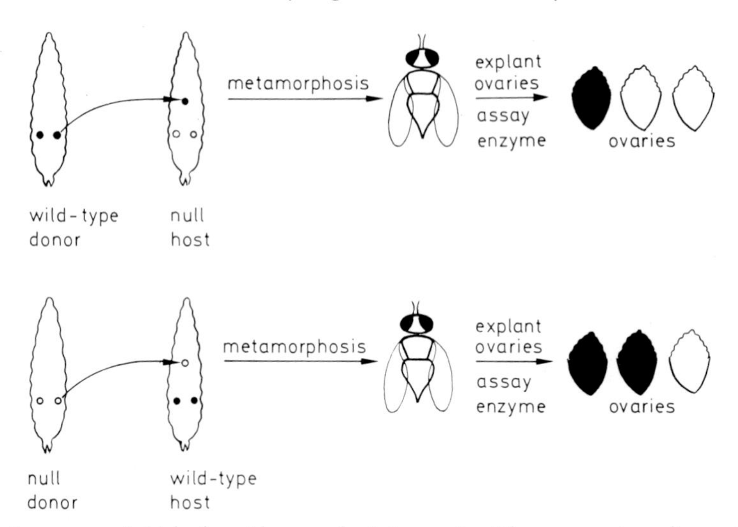

Fig. 20. Autonomy of aldehyde oxidase synthesis in ovaries. The ovary primordia are indicated as *Aldox*$^{+}$ (solid) or *Aldox*^{n} (null mutant, hollow). Three ovaries are recovered from each host. They are indicated as containing aldehyde oxidase activity (shaded) or no activity (unshaded). The results indicate that in each case, the implanted ovary behaved according to its own genotype

negative cells in Malpighian tubule and gut. This suggests autonomy of synthesis at the cellular level, i.e., the aldehyde oxidase found in a given cell was made there. Unfortunately, similar information on XDH is not yet available.

There is considerable evidence that suggests regulation after translation may be important in this system of enzymes. First of all, it seems almost certain that the products of the enzyme's structural genes interact with the products of the *ma-l* and/or *lxd* loci after synthesis. This provides the possibility of control through that product level interaction. More directly, there is strong evidence that XDH, and perhaps the other two enzymes, can exist in alternate forms with different catalytic efficiencies (COLLINS *et al.*, 1970; GLASSMAN *et al.*, 1968).

Flies grown on a synthetic medium rich in protein show a 4–7-fold increase in XDH activity over flies on standard cornmeal-agar medium. Pyridoxal oxidase is similarly increased, while aldehyde oxidase shows a much smaller increase in activity. Kynurenine formamidase, an unrelated enzyme, showed no such increase. The same response is found in strains with different electrophoretic forms of XDH. Mutant *ry* and *ma-l* flies formed no XDH in the synthetic medium, and *lxd* mutants have the usual low level. Electrophoretically, the enzyme produced by flies on synthetic medium is distinguishable from the usual form. The XDH band is broader, and migrates faster toward the anode. The two forms of XDH found here do not, however, seem to correspond to the alternate forms generated by various treatments *in vitro* (XDH-I and XDH-II referred to in the biochemistry section). The XDH extracted from flies grown on synthetic medium is not separable from normal XDH on DEAE-cellulose, while XDH-I and XDH-II are separable.

The increase in enzyme activity does not seem to be accompanied by a corresponding increase in CRM. This, and the altered electrophoretic mobility, seem to favor a model of increased enzyme activity through greater catalytic activity per molecule present, rather than an increase in the number of molecules.

DICKINSON (1969) has found changes in the ratio of aldehyde oxidase activity to CRM during development. Larvae contain about a two-fold excess of CRM. This observation was confirmed by HERDING (1970), who also found that crude extract of larvae showed a 50–100% increase in enzyme activity when incubated briefly at 55–60°. Adult extracts show no such activation. These observations suggest that larvae either contain a pool of inactive precursor molecules that can be heat activated, or that they contain a less active form of the enzyme that can be converted to the more active form by heating. In either case, the phenomenon provides an opportunity for regulation of enzyme activity beyond the translational level.

Developmental History of Uricase

Although not really a member of this system of enzymes, uricase is considered here because it acts on a product of XDH, and the effect absence of XDH has on uricase has been examined. FRIEDMAN (1973) found decreasing levels of uricase in larvae, and essentially no activity in pupae. In wild-type flies, a modest increase was found following eclosion. In *ry* and *ma-l* this increase was dramatic. Since *ry* and *ma-l* flies lack uric acid, this molecule cannot be an inducer of uricase. However, the abnormally high level of uricase in *ry* and *ma-l* could indicate that some precursor which accumulates in the mutants stimulates uricase activity in some way.

Summary

Although the system of genes and enzymes discussed in this chapter is complex and incompletely understood, several reasonably firm conclusions do emerge.

1. Despite the number of genes and enzymes involved, there is no real support for an operon-like model of this system.

2. As with most other enzymes in *Drosophila*, the strict proportionality between normal structural genes present and enzyme activity suggests that there is no feedback control of enzyme level, but rather, the enzyme is regulated by a system that is "blind" to the actual amount of enzyme being produced.

3. Nevertheless, the enzymes (at least aldehyde oxidase and probably XDH), are very precisely regulated during development by some other mechanism.

4. Regulation at the level of product interaction and conformational change are very real possibilities.

Dehydrogenases

Several members of the class of enzymes referred to as dehydrogenases have been studied in *Drosophila*. These enzymes have been particularly well suited for biochemical and genetic studies because of the availability of a rapid, sensitive, quantitative assay, and a histochemical method of detection that has been readily adapted to electrophoretic supporting media allowing for the screening of electrophoretic variants (Fig. 21). The quantitative assay is dependent on measuring the changes in absorbance at 340 nm of the appropriate pyridine nucleotide, either NADH or NADPH. The reaction generally may be follwed in either direction by recording increases or decreases in the concentration of reduced pyridine nucleotide. The assays in electrophoretic supporting media are performed by incubating the gels in a solution containing appropriate substrate and the oxidized form of the coenzyme. The reduced NADH or NADPH formed during the reaction pass electrons through an intermediate electron carrier, usually phenazine methosulfate, to a tetrazolium compound resulting in the formation of an insoluble, purple diformazan dye at the site of enzyme activity. To date, twelve dehydrogenase from *Drosophila* have been investigated. Of these, electrophoretic variants have been found

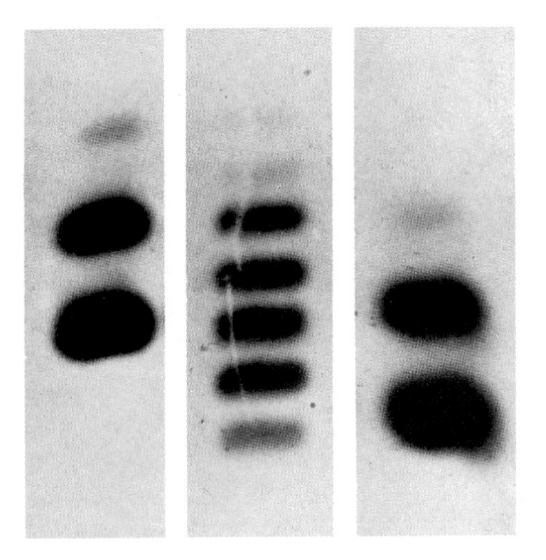

Fig. 21. Representative electrophoretic gel stained for alcohol dehydrogenase. Extracts of the two parental strains are on the sides, and an extract of F_1 hybrids is in the middle. (From URSPRUNG and LEONE, 1965)

for seven. There have been no reports of screening for two others and one, lactate dehydrogenase, has been the subject of a search for variants, but none have been found. Two others are mitochondrial enzymes and present technical problems.

Alcohol Dehydrogenase (EC 1.1.1.1)

Biochemistry of Alcohol Dehydrogenase (ADH)

ADH catalyses the oxidation of an alcohol and the concurrent reduction of NAD to yield the appropriate aldehyde and reduced NAD (NADH). A variety of alcohols may serve as substrates for *Drosophila* ADH. Among them are ethanol, propanol, isopropanol, butanol, 2-butanol. octanol, glycerol and cyclohexanol. SOFER and URSPRUNG (1968) have reported that *Drosophila* ADH undergoes substrate activation at high concentrations of 2-butanol, 2-propanol and cyclohexanol. Substrate activation is not observed using primary alcohols as substrates. JACOBSEN *et. al.* (1970) have also reported increased rates using secondary alcohols. *Drosophila* ADH apparently shows no activity toward methanol (JOHNSON and DENNISTON, 1964; JACOBSON *et al.*, 1970).

The function that ADH serves in *Drosophila* is not clear. It is not vitally essential to fhe fly, since GRELL *et al.* (1968) have isolated mutants that have no ADH. PIPKIN (1968) has also found *Drosophila* in natural populations that are lacking ADH.

Isozymes of ADH

The first descriptions of ADH from *Drosophila* revealed that this enzyme exists in several forms, as indicated by different mobilities during electrophoresis (JOHNSON and DENNISTON, 1964; GRELL *et al.*, 1965; URSPRUNG and LEONE, 1965). The number of ADH isozymes observed by various workers has ranged from two to ten. The resolution of this complex picture has served to highlight several of the many factors that can result in more than one band of activity being seen after the staining of gels subjected to electrophoresis. JOHNSON and DENNISTON (1964) originally observed two bands of ADH activity in extracts of single flies. There was one major band and one faint band that migrated faster towards the anode. Genetic differences in electrophoretic mobilities were also seen. Some strains possessed "fast" migrating ADH patterns, and some had "slow" patterns. Heterozygous individuals obtained by crossing flies of each genotype yielded zymograms containing five bands. The locus responsible for these electrophoretic variants has been called the *Adh* Locus. These observations were extended by GRELL *et al.* (1965) and URSPRUNG and LEONE (1965). Using a somewhat more sensitive modification of the assay, three bands could be observed in single flies from inbred strains. From crosses between flies having the "slow" ADH pattern and flies with the "fast" pattern, GRELL *et al.* obtained flies having zymograms with 9 bands, the 6 parental bands plus three hybrid bands. URSPRUNG and LEONE (1965) observed seven activity bands in the heterozygous individuals. Additional, slower moving bands having ADH activity can also be seen in the electrophorograms of URSPRUNG and LEONE. These bands show increased activity using longer chain alcohols as substrates. COURTRIGHT *et al.* (1966) have shown that these additional bands, while having some ADH activity, are both genetically and biochemically a separate enzyme system form ADH. This enzyme has been called octanol dehydrogenase (ODH). Therefore, one can disregard the three ODH brands; but there still appear to be seven activity bands which must be accounted for to gain a complete understanding of the structure and genetics of ADH.

Since the variants of ADH show variation in the entire zymogram pattern, presumably all isozymes must have at least one polypeptide chain from the same locus. Since in genetic hybrids, hybrid enzyme molecules are observed, there must be more than one polypeptide chain per ADH molecule. Several lines of evidence have indicated that the isozymes of ADH do not represent different-size aggregates of a single protein species (GRELL *et al.*, 1968; IMBERSKI *et al.*, 1968; JACOBSON *et al.*, 1970). Two general types of models of ADH structure have been proposed to account for the isozyme patterns and the above mentioned properties. One of these postulates that the different electrophoretic mobilities are due to combinations of polypeptides from the *Adh* locus with polypeptides from other, as yet, unidentified loci. The second model proposes that the isozyme pattern results from differential modification of the protein after synthesis, adding to or altering some propertie(s) that affects the charge of the molecule. It should be pointed out that the possibility of epigenetic modification does not exclude the possibility that ADH may be a heteropolymer having different types of subunits. It only implies that the charge differences are due to the post-synthetic modification. Therefore the demonstration of epigenetic modification would not prove that the molecule is a homopolymer, and the product of only a structural gene. Demonstration of this requires an independent verification that the molecule is composed of only one type of polypeptide chain.

The initial indication that post synthesis modification played a role in ADH isozyme formation came from the observation of URSPRUNG and CARLIN (1968), who found that the isozyme pattern could be altered *in vitro* by a variety of treatments. These included dialysis against buffer containing 2-mercapoethanol and NADH, buffer containing only 2-mercaptoethanol, or even prolonged storage in the cold. These alterations in pattern were of two types. There were alterations in relative staining intensities of original bands, and the production of new bands not usually observed in the original extracts. A clue as to the nature of this modification came from experiments in which extracts containing ADH were treated with NAD^+ prior to electrophoresis. The zymograms from NAD^+-treated extracts showed an increase in the relative proportion of the more negatively charged isozymes. This observation was extended by performing electrophoresis in gels that were formed containing NAD^+. When sufficiently high concentrations of NAD^+ were used, the resulting zymograms contained only one darkly staining band. The appearance of only one band of ADH in conditions apparently saturating the enzyme with coenzyme suggested strongly that the multiple ADH bands were the result of different degrees of saturation of the enzyme with coenzyme. Under saturating NAD^+ conditions the electrophoretic distinction between genetic variants remained. Each strain had only one band, and these bands had different mobilities.

JACOBSEN (1968) has also obtained evidence that indicates that the different isozymes may be the result of different amounts of bound NAD^+ on a single protein species. ADH has been purified, using standard procedures, to the point that it sediments as a single peak in the analytical ultracentrifuge, and migrates as a single protein band during electrophoresis in polyacrylamide gels. This purified ADH shows only one activity band in gels, and migrates as ADH-5 as compared to position of zymograms of crude extracts. When purified ADH is incubated with NAD^+, additional, more anodally migrating isozymes equivalent in mobility to ADH-3 and ADH-1 can be produced. JACOBSON (1968) has eluted these experimentally produced isozymes from the gels and determined that they do have substantially more NAD^+ bound to them than does the original purified ADH from which they were derived. These observations led to the apparently reasonable hypothesis that the isozyme bands of ADH seen in extracts were generated by a single enzyme species with different amounts of bound coenzyme, resulting in differently charged molecules. However, recent investigations by JACOBSON and coworkers have indicated that co-enzyme binding may not be causally related to the isozyme pattern. JACOBSON *et al.* (1972) have used the *in vitro* generation of ADH-1 from purified ADH-5 as a model system to study the basis for the isozyme pattern obtained from whole animal extracts. The conversion of ADH-5 to ADH-1 can be effected by NAD^+. The rate of this conversion is dependent on enzyme and NAD^+ concentration. Furthermore, bound NAD^+, in varying amounts, can be found on the experimentally produced ADH-1. However, the amount of NAD^+ remaining after conversion depends on the conditions of dialysis by which excess NAD^+ is removed. The ADH-1 produced has the same electrophoretic mobility, whether it retains lower or relatively high levels of bound NAD^+. Therefore the change in electrophoretic mobility is not directly related to the amount of NAD that is bound. In addition it was discovered that acetone can also cause the conversion of ADH-5 to ADH-1, even in the absence of added NAD^+. From these results it is apparent that the difference between ADH-5 and ADH-1 involves something other than coenzyme binding. JACOBSON *et al.* (1972) propose that conformational changes are involved in the conversion, which result in a molecule with a different net charge. To establish this, KNOPP and JACOBSON (1972) have conducted fluorescence studies on ADH. During the conversion of ADH-5 to ADH-1 they observed changes in the fluorescence emission maximum to a shorter wave length, and a decrease in the relative quantum yield. Analysis of the fluorescence spectra has indicated that this change in quantum yield is due to alterations involving tryptophan residues. Therefore the environment of the tryptophan residues is changed when ADH-5 is converted to ADH-1. This demonstrates some difference in the secondary or tertiary structure of the two isozymes. This difference could result in the observed difference in electrophoretic mobility. It is not yet certain whether the isozyme bands obtained on gels do represent conformational isomers. This appears the most likely hypothesis currently available, but it depends on whether the ADH-1 generated *in vitro* is the same as the native ADH-1. Some differences in their properties have been noted (JACOBSON *et al.*, 1972).

Structure of ADH

SOFER and URSPRUNG (1968) have developed procedures for purifiying ADH. These preparations show coincident electrophoretic migration of a single protein species and ADH activity bands. They also show a single peak in sedimentation velocity experiments conducted in the analytical ultracentrifuge. The molecular weight of ADH has been estimated to be 4.4×10^4 by calculation from a sedimentation coefficient of 3.9, which was obtained from sedimentation velocity in sucrose gradients and in the analytical ultracentrifuge.

JACOBSON *et al.* (1970) have also purified *Drosophila* ADH, and have subjected it to amino acid analyses. There are present 131 amino acid residues which represent a minimum molecular weight of 13960. There is approximately one tryptophan and 1 cysteine per M.W. of 15000. There is no methionine, and since the total moles of sulfhydryl is approximately equal to the moles of cysteine, there does not appear to be any disulfide present. Preparations of purified ADH with only one electrophoretic band of ADH activity were obtained by JACOBSON *et al.* (1970). This is ADH-5, which is the relatively most electropositive of the isozymes. ADH-5 was converted to ADH-1, the most electronegative isozyme by incubation with NAD. Electrophoresis of this resultant ADH-1 sample showed that all protein in the sample, as well as all enzyme activity, had the altered electrophoretic mobility. This provides additional, quite convincing, criteria for the purity of the original ADH-5 preparation.

JACOBSON and PFUDERER (1970) have determined the molecular weight of ADH by equilibrium centrifugation and gel filtration chromatography. By the former method the determined value was 60200, and by gel filtration it was 58000. These values are somewhat higher than the value of 44000 reported by SOFER and URSPRUNG (1968). JACOBSON and PFUDERER have investigated the subunit structure of ADH by observing changes in size resulting from exposure to a variety of dissociating agents including urea, maleic anhydride and sodium dodecylsulfate (SDS). A variety of different-sized dissociation products have been obtained. The smallest of these has an approximate molecular weight of 7400. There is some difficulty in drawing firm conclusions about subunit structure from these observations. The genetic data indicate only one kind of subunit. The amino acid composition would place the minimum molecular weight of this subunit in the range of 15000, since there is one tryptophan and one cysteine for a peptide of this size. A size of 7400 would require at least two different types of subunits, one containing tryptophan and one not containing tryptophan, and one containing cysteine and one not containing cysteine. From their dissociation studies JACOBSON and PFUDERER would like to conclude that ADH is an octamer with different types of subunits. However, they do suggest that the 7400 subunit could be produced by cleaving a single chain to yield 2 different chains. In this case ADH would be considered a tetramer, composed of identical subunits.

Genetics of Alcohol Dehydrogenase

GRELL *et al.* (1965) have utilized the electrophoretic variants of ADH to map the *Adh* locus. This locus has been positioned on the second chromosome, approximately one tenth the distance from *elbow (el)* at 2:50.0 to *reduced-scraggly* (rd^s) at 2:51.2. It has therefore been assigned the position of 2:50.1. This map position has been confirmed by COURTRIGHT *et al.* (1966).

GRELL *et al.* (1968) have succeeded in inducing a third allele at the *Adh* locus by ethyl methanesulfonate mutagenesis. This allele, referred to as Adh^D, is responsible for the production of an ADH with an increased electronegativity. ADH from the Adh^D locus migrates faster than ADH produced by the Adh^S or Adh^F loci. This variant has also been mapped to 2:50.1. In heterozygous individuals with Adh^D/Adh^F or Adh^D/Adh^S, the appropriate hybrid enzyme bands are observed as well as the parental bands. This demonstrates that the induced mutant Adh^D is truly a mutant in the same gene coding for the ADH^S and ADH^F enzymes. The induction of an electrophoretic variant should prove very useful in the genetic analysis of enzymes in which no naturally occurring variants can be found. However, it should be noted that this may require a rather formidable effort. GRELL *et al.* (1968) scored 5800 flies, and found one induced variant.

These authors then used an additional approach that may make finding mutants of a particular enzyme more possible when the locus is already known. They have produced a deficiency which covers the *Adh* locus. The deficiency has been localized on the salivary chromosome map, and extends between 34E3-34F1 and 35C3-35D1. This deficiency was obtained by X-irradiation of males which carried the Adh^S gene. These males were then crossed to females carrying the Adh^F locus, and a series of recessive marker mutations linked to the *Adh* locus. Progeny of this cross examined for the presence of flies showed more than one of the marker mutations. Since the probability of coincident mutation at two of these loci is low, those flies showing more than one of the marker genes originally coming from the female parent were suspected to carry deficiencies in the chromosome contributed by the male parent. To establish that such suspected deficiencies covered the *Adh* locus, these flies were crossed to flies carrying the Adh^F allele. If the chromosome in question really was deficient for the *Adh* locus, the progeny of this cross should only show ADH^F. On the other hand, if flies were not deficient for *Adh*, then their progeny should show the typical hybrid pattern expected in Adh^S/Adh^F heterozygotes. GRELL *et al.* (1968) obtained flies which carried a deficiency by such criteria, and confirmed the deficiency as reflected by a loop in salivary chromosomes. The existence of this deficiency for the *Adh* locus has been used by GRELL *et al.* (1968) to facilitate the isolation of flies with mutant alleles at the *Adh* locus. Male progeny were obtained from a cross of EMS treated males, and females carrying appropriate chromosomal markers. These were mated to females heterozygous for the *Adh* deficiency. Progeny of this cross are "haploid" for the *Adh* locus, and therefore any recessive mutations induced by the EMS treatment will be evident without the necessity of further crossing to establish homozygosity. Several types of mutants at *Adh* have been isolated by this procedure. These include flies showing a complete lack of enzyme, and some with altered enzyme. The absence of ADH can be detected using the standard ADH assay. A simpler selection method was devised by GRELL *et al.* to detect ADH deficient flies. This approach takes advantage of the fact that ADH negative flies are sensitive to ethanol in the medium, and when placed on a medium containing 15% ethanol first lose control of motor activities and then die within 24 hrs. This has provided a useful screening procedure for selecting mutants from crosses of EMS treated males and females carrying the deficiency for *Adh*. Several types of mutants have been isolated. Adh^{n2}, Adh^{n3}, and Adh^{n4} are loci causing the lack of enzyme activity, and also do not form hybrid bands when crossed to flies carrying other *Adh* alleles. Therefore these mutations may be completely deficient in any product from the *Adh* locus. Adh^{n5} has very low levels of ADH activity, and when crossed to flies carrying Adh^{F} yield progeny whose zymograms show active bands corresponding to the hybrid type enzyme molecule. Therefore it appears that Adh^{n5} produces an altered partially active protein molecule, deficient in catalytic activity but still able to form reasonably normal subunit associations. Adh^{n1} is similar to Adh^{n5} in that it produces hybrid enzyme molecules when heterozygous with other alleles, but when homozygous no enzyme activity can be detected. The procedures developed by GRELL *et al.* (1968) have resulted in the *Adh* locus probably being the most extensively studied locus with respect to its enzyme product. It is to be hoped that these general procedures will become applicable to other gene-enzyme systems in the near future.

SOFER and HATKOFF (1972) have devised an effective positive selection system for ADH negative mutants. The secondary alcohol 1-pentene-3-ol (pentenol) is poisonous to wild-type *Drosophila*. Presumably it is oxidized to a toxic ketone by ADH, since known ADH negative mutants are resistant while all ADH positive strains that were tested are sensitive. SOFER and HATKOFF have demonstrated that pentenol is a substrate for *Drosophila* ADH. These authors treated 100000 flies which were prospective ADH negative mutants as a result of treatment with EMS. Twelve new ADH negative mutants were obtained with only two false mutant survivors. This screen appears to be very powerful and effective, and should prove quite useful in obtaining a complete description of the genetic elements related to ADH.

Developmental Biology of Alcohol Dehydrogenase

The activity of ADH during development has been determined by URSPRUNG *et al.* (1968). Starting at low levels in embryos, the specific activity begins to rise just before hatching. This increase continues during the larval stages, and reaches a maximum in late third instar. Subsequent to puparium formation, the activity gradually declines to intermediate values which remain for most of the duration of the pupal stage. At emergence, specific activity increases approximately two-fold, to reach a second maximum in adult flies. DUNN *et al.* (1969) have reported that the activity continues to rise after emergence until the fourth or fifth day. Levels remain constant thereafter until a decline begins on about day twelve. These data are summarized in Fig. 22. WRIGHT and SHAW (1970) have investigated more closely the initial rise in activity which occurs late in embryogenesis. Individuals heterozygous for the alleles Adh^{F} and Adh^{S} show the zymogram characteristic of the maternal parent during the first eighteen hours after fertilization. Evidence of the paternal form of the enzyme is first seen at 22–24 hrs of development. This time is coincident with the initial increase in specific activity. These data suggest that gene(s) responsible for ADH synthesis may undergo initial activation at 22–24 hrs post fertilization.

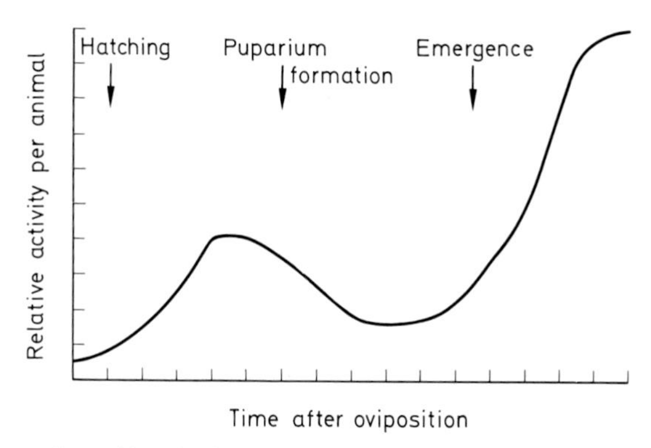

Fig. 22. Developmental profile of alcohol dehydrogenase. (Data redrawn from URSPRUNG *et al.*, (1968)

ADH is found in several tissues in larvae but the major portion, 40–50%, is found in the fat body. Lesser amounts are found in the intestine and Malphigian tubules. There is no detectable ADH in larval brain, salivary glands or imaginal discs (URSPRUNG *et al.*, 1970). It is interesting to note that the highest specific activity in the fat body occurs during second instar. However, due to the large amount of growth of the fat body in third instar, the highest total activity per fat body occurs in late third instar. ADH is one of several enzymes that has been found in the larval integument by KNOWLES and FRISTROM (1967).

URSPRUNG *et al.* (1970) have also examined the tissue distribution of ADH in adult flies. There is substantial activity in the head and fat body, lower levels are found in the Malphigian tubules and intestine and some parts of the genital apparatus. The activity in the genital apparatus is of special interest. ADH was detected in the vasa deferentia, paragonia, ductus, sperm pump muscles and rectum. Since there is no activity in the imaginal discs, presumably ADH found in these genital disc derivatives must be produced during the differentiation of the disc cells. This has been confirmed by transplanting genital discs to hosts which genotypically were of the ADH negative type. Upon differentiation of these transplant discs, ADH activity was observed, establishing that the genital apparatus ADH was produced by the differentiating discs themselves, and not transported from some other part of the body.

Experiments performed by HORIKAWA *et al.* (1967) have indicated that the levels of ADH may be affected by external sources of ethanol. Embryonic cells when cultured in media supplemented with various concentrations of ethanol responded by showing increases in ADH activity over a period of 42 hrs. Several aspects of this response make a simple interpretation of these results difficult in terms of any mechanism which could be analogous to enzyme induction. The magnitude of the response was not ethanol dependent. Also similar responses were obtained using methanol, which is not a substrate for the enzyme, while no response was obtained when isopropanol, a known substrate, was used. The increases in specific activity observed were dependent on a continuous supply of ethanol in the medium. It should be noted that this response was specific for ADH in that other enzymes, XDH and alkaline phosphatase, did not show increases in specific activity in response to ethanol. It seems possible that some sort of enzyme stabilization may be effected by the ethanol that may or may not be related to its role as a substrate. However, the use of cultured embryonic cells would appear to be a most promising area to proceed to investigate the regulation of enzyme activity in *Drosophila*.

The extent of biochemical, genetic and development studies already conducted on ADH make this enzyme system probably the most broadly characterized gene-enzyme system in *Drosophila*. Therefore, it appears to be a most promising one in which to investigate the regulatory mechanisms which operate to control enzyme levels during development.

Octanol Dehydrogenase

Octanol dehydrogenase (ODH) is an enzyme similar to ADH, but with different substrate specificities. This enzyme was first described by URSPRUNG and LEONE (1965). It is detected on gels after electrophoresis, using longer chain alcohols such as l-octanol or benzyl alcohol as substrates in the standard dehydrogenase assay.

Biochemistry of ODH

Several lines of evidence demonstrate that ODH and ADH are separate enzymes. COURTRIGHT *et al.* (1966) have shown several differences in their biochemical properties. These include substantial differences in electrophoretic mobilities and separation on DEAE cellulose ion exchange chromatography. Partial purification of ODH has been reported by SIEBER *et al.* (1972). By combining ammonium sulfate fractionation and chromatography on DEAE-Sephadex and hydroxylapatite, a 75-fold purified preparation has been obtained in a yield of 15.5% over the starting homogenate. The molecular weight of ODH was estimated by gel filtration to be 109000. Substrate specificity studies revealed that ODH can utilize several primary, straight chain alcohols. Affinity for substrate increases with chain length. A variety of secondary and branched chain alcohols were also tested, and showed no activity. In addition, benzyl alcohol and farnesol can be utilized as substrates (SIEBER *et al.*, 1972). MADHAVAN *et al.* (1973) have suggested that ODH may have a role in juvenile hormone catabolism. The utilization of farnesol and the developmental profile of ODH are consistent with this hypothesis.

D. melanogaster homozygous at the ODH locus show only one band upon electrophoretic analysis, and heterozygotes for the two *Odh* alleles show three. PIPKIN (1968, 1969), has investigated electrophoretic patterns of ODH shown by various other species including *D. metzii* and *D. pellewae*. Isozyme patterns from these species can be quite complex. PIPKIN has indicated that isozymes may be found at 13 positions in agar gels after electrophoresis. Individual flies may show as many as nine bands or as few as one. The inheritance of ODH isozyme patterns in crosses between different strains has been studied in order to determine the number of structural genes involved in ODH synthesis, and also to estimate the number of subunits in the ODH molecule. In one cross, backcross progeny were observed to segregate for four different patterns in a 1:1:1:1 ratio. This suggests that the ODH patterns depend on at least two structural genes. In order to account for these results and other bands seen in different crosses, PIPKIN and BREMNER (1970) have suggested that ODH is a tetramer, and that three structural genes may be involved in its synthesis.

Genetics of ODH

Electrophoretic variants of ODH are available, and these have been used by COURTRIGHT *et al.* (1966) to map the *Odh* locus at 3.49.1. Therefore *Odh* and *Adh* are genetically separable, since *Adh* maps at 2.50.1.

OGANJI (1971) has recently investigated ODH zymogram patterns which are obtained using extracts from several strains of *D. albirostris* and progeny from crosses between these strains. Flies from this species show a multibanded ODH zymogram which is interpreted as indicative of a high degree of polymorphism. Crosses between true breeding individuals of isolated strains yield zymograms with three or more ODH bands. Analysis of the genetic data indicated that the ODH pattern is the result of monofactorial inheritance, although some exceptional cases were noted. OGANJI favors a model having two structural genes for ODH and a tetramer structure for the enzyme, to account for the several bands that are observed. Using this model OGANJI invokes the existence of a regulatory gene to explain some exceptional cases. The regulatory function supposedly affects either the time or rate at which one or the other of the structural genes is translated. Regulatory genes were also invoked to explain ODH zymograms by PIPKIN (1968, 1969) and PIPKIN and BREMNER (1970). The postulation, based on only zymogram patterns, of a regulatory gene is premature. As OGANJI points out, the kind of analyses which have been done are insufficient to understand the subunit structure of an enzyme. Since no biochemical information is available concerning the structure of the enzyme or factors which can affect its electrophoretic mobility, utmost caution should be exercised in forming hypotheses regarding subunit structure, the number of structural genes or the existence of regulatory genes. Presumably these situations will be clarified when biochemical information becomes available concerning the basis of the electrophoretic differences between the isozymes.

Developmental Biology of ODH

Octanol dehydrogenase has been characterized developmentally by MADHAVAN *et al.* (1973). The activity per organism is depicted in Fig. 23. After each molt, activity rises. This post-molt rise is more evident when specific enzyme activities are plotted as a function of development. ODH has been found in all tissues tested with particularly high activity seen in the *corpus allatum* region of the ring gland. The observations are used by the authors to suggest a possible role in juvenile hormone metabolism.

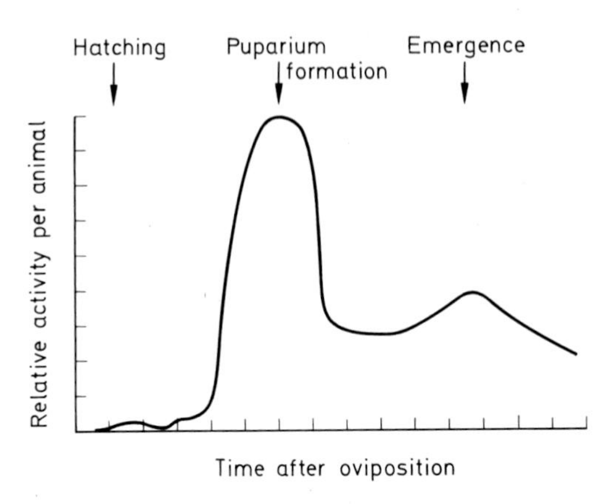

Fig. 23. Developmental profile of octanol dehydrogenase. (Redrawn from MADHAVAN *et al.*, 1972)

Developmental variation in the ODH isozyme pattern of *D. pellewae, D. metzii, D. leticiae* and some intraspecific hybrids of these has been reported by PIPKIN and BREMNER (1970). There appears to be a progressive loss of certain isozymes, such that in third instar larvae only one band may be observed in the appropriate genetic situation. The exact pattern of this isozyme shift is strain specific. In some strains faster moving isozymes are lost and slower bands predominate in adults, while in other strains, slower bands disappear and fast bands remain in the adult. Since the molecular basis of these ODH isozymes is unclear at this time, it is not certain whether this developmental shift is the result of differential activity of one or more genes coding for ODH, or whether it represents a change in conditions which effect some epigenetic modification of the ODH molecule.

Glucose-6-Phosphate Dehydrogenase (EC 1.1.1.49)

Biochemistry of G6PD

Glucose-6-phosphate dehydrogenase (G6PD) from *Drosophila* was first described by YOUNG *et al.* (1964). The enzyme has been observed to occur in different electrophoretic forms. Nineteen strains were examined by YOUNG, and five were found to contain a single slow moving band (B). Twelve contained a single faster moving band (A). Two strains were found to be polymorphic, having both A and B. Heterozygous individuals obtained by crossing B-type flies with A-type flies show only two bands of G6PD activity which correspond to the parental positions. The molecular basis for the mobility difference of the two forms of G6PD has been investigated by STEELE *et al.* (1969). They have partially purified both A and B forms by a succession of DEAE ion exchange chromatography, ammonium sulfate fractionation, CM ion exchange chromatography and column chromatography on hydroxylapatite. Form A obtained by these procedures was 242-fold purified, and was obtained in a yield of 38%. The behavior of forms A and B during these procedures was identical, however, the yields of form B were lower. The molecular weight of both forms has been estimated by using eight-channel starch gel electrophoresis. Form B has an approximate molecular weight of 317000, form A 147000. This suggests that form B may be formed from two molecules having the size of form A. This size-difference has been confirmed by measurements of STOKE'S radii using gel filtration chromatography. The STOKE'S radius of form B was found to be 61.5 Å, and that of form a, 47.6 Å. Therefore it appears that the basis for the two forms during electrophoresis is failure of form A to polymerize properly, thereby yielding a molecule with half the size of form B.

STEELE *et al.* (1969) have argued further that size is the only difference between the two forms. In flies having form B, a fainter band is often seen at the A position. These sub-bands may result from partial dissociation. Also, form B may be experimentally dissociated by raising the ionic strength above 0.1 M and lowering the pH below 7.5. This treatment leads to an increase in enzyme migrating to the A position. The minor, naturally occurring sub-bands and the band resulting from dissociation, appears to have the same size as the naturally occurring A variant and the A band seen in flies heterozygous for the two forms, as judged by electrophoresis. Therefore, since all these "fast" bands migrate to the same position, and since they have the same size, they must have the same charge. STEELE *et al.* (1969) have concluded that since the only difference observable between forms A and B are in size, the structural basis for the enzyme polymorphism must be the result of a change in an amino acid which does not contribute to the net charge of the molecule, but plays a role in subunit association. This argument assumes that the mutation is in the structural gene for G6PD, and not in a gene which functions to produce a factor involved in subunit association. These workers have attempted to effect association of the A-variant to form B, but this has not yet been possible.

The enzymatic characteristics of the two forms have also been compared. STEELE *et al.* (1969) determined the K_m for NADP of the B form to be 5.8×10^{-5} M, and for the A form to be 2.5×10^{-5} M. The K_m for glucose-6-phosphate of the B form is 1.7×10^{-4} M, and of the A form 3×10^{-4} M. These results suggest a catalytic difference in the two forms. KOMMA (1968) has determined the K_m for glucose-6-phosphate of form A from males to be 5.6×10^{-4}, and from females 3.5×10^{-4}. G6PD from males having form B shows a K_m for glucose-6-phosphate of 5.4×10^{-4}, while in females with form B the K_m was found to be 3.6×10^{-4} M. Therefore, there appears to be a sex difference in K_m, but the results seem to indicate that the K_m's with respect to G6PD for either form from the same sex are similar. This apparent difference between these results of KOMMA (1968) and those of STEELE *et al.* (1969) may be due to the difference in pH of these respective assays. KOMMA performs assays at 7.5, while STEELE *et al.* use pH 8.6. As previously mentioned, lower pH favors the dissociation of form B to form A. Therefore, at pH 7.5 or below it would be expected that the assays contained a large proportion of dissociated B. This might tend to minimize the appearance of any catalytic differences between the two isozyme forms. At the higher pH of 8.6 there is little dissociation of form B, and therefore one would expect assays under these conditions to maximize any differences in catalytic properties that might exist between the two isozymes. The basis for the K_m differences between males and females is unclear at this time, but correlates with other differences of the enzyme from males and females reported by KOMMA (see below). Differences between form A and form B in their stability upon heating have been reported by STEELE *et al.* (1969). They observed the rate of loss of activity at 44 °C in extracts from flies with form A to be greater than extracts containing form B. Extracts of individuals heterozygous for A and B showed an intermediate rate of loss of activity. *In vitro* mixtures of forms A and B also showed intermediate thermal stability. KOMMA (1968) has observed that activity in extracts from B-type flies is slightly more stable at 50 °C. G6PD is usually unstable in the absence of NADP. This co-factor, therefore, is usually included in homogenizing buffers and other solutions containing enzyme activity. KOMMA has reported that the B-form is less sensitive to the absence of NADP than is the A-form, provided that the extracts are maintained with 10^{-2} M EDTA.

Genetics of G6PD

The genes responsible for the electrophoretic variation of G6PD are located on the X-chromosome. YOUNG *et al.* (1964) observed that females may be phenotypcially form A, form B or have both bands. Males, on the other hand, are either form A or form B, the two-band heterozygous condition is not found. A cross of A-type males with B females yields male progeny of type B, and females having both bands. The reciprocal cross yields males of type A and females with both A and B bands. This establishes the sex-linked, co-dominant mode of inheritance for G6PD isozymes. The gene responsible for the electrophoretic variation of G6PD has been mapped by T. WRIGHT (quoted in SEECOF *et al.*, 1969) to position 62.83 of the X-chromosome. This locus has been designated as *Zw*.

The discovery of the sex-linked mode of inheritance has led several investigators to attempt to study the problem of dosage compensation at the molecular level. KOMMA (1966) SEECOF *et al.* (1969) and STEELE *et al.* (1969) have all reported that G6PD activity in *Drosophila* is dosage compensated, i.e. males and females of the same strain have the same levels of activity, even though the female has two doses of the *Zw* locus and the male only one. In one case, STEELE *et al.* (1969), males were reported to have a somewhat higher specific activity than females. This apparent anomaly is due to the presence of G6PD-deficient eggs in the females, which contribute substantially to the amount of protein and thereby reduce the specific activity. In the combined head-thorax parts of the body, the specific activities of males and females are found to be similar. SEECOF *et al.* (1969) have created stocks which carry different numbers of the G6PD gene. Females were constructed which have a region of the X-chromosome containing the *Zw* locus translocated to the fourth chromosome. These females have three doses of the *Zw* locus, one on each X and one on chromosome four. Such females show an increase of G6PD activity to 1.39 times that of their normal two-dose sisters. On the other hand, triploid females have levels of activity equivalent to that of diploid females. Males which carry the translocation have activity levels which are 1.52 times that of their normal brothers. SEECOF *et al.* have proposed a model for dosage compensation to account for this regulation. The model postulates both an autosomal factor which functions to promote activity of the compensated gene, and also an X-linked regulatory element. The existence of both factors appears necessary to account for G6PD levels in the various genotypes.

While the specific details of this model await experimental verification, it is clear that regulation in some form must be occurring. It would be most interesting to extend these observations to a measurement of the actual rate of synthesis of this enzyme or another enzyme that has its structural gene on the X-chromosome.

KOMMA (1966) has studied the effect of the genes *transformer (tra)* and *doublesex (dsx)* on the G6PD levels. Females homozygous for *tra* appear as males, both internally and externally, but they are sterile. These pseudomales have somewhat higher specific activities of G6PD than their sisters which are heterozygous for *tra*. This effect is more pronounced at the lower growth temperature of 21° C. The levels of G6PD in pseudomales and comparable females are equivalent when these are raised at 29° C. This difference in specific activity observed between *tra/tra* XX pseudomales and +/*tra* females is caused by a lowering of the specific activity of the control females. That is, when these flies are compared to their normal brothers or to other flies carrying the same G6PD alleles, the levels of activity of the pseudomales is approximately the same. However, the control females have a lower specific activity. These control females contain one third chromosome carrying *tra*, and one carrying *Dichaete (D)*. KOMMA has postulated the existence of a recessive third chromosome gene, but not *tra* itself, which affects G6PD activity and happens to be carried on both the *D* and *tra* carrying third chromosomes. However, STEELE *et al.* (1969) have pointed out that the difference in specific activity between pseudomales and females may be due to the presence of G6PD-deficient eggs in the females. It would be interesting to see the results of KOMMA expressed on a units per animal basis. This mode of expression would give a partial answer to this question. The gene *dsx*, which causes the production of phenotypic intersexes from both males and females, was also tested for its effect on G6PD activity. Intersexes of the XXY/*dsx* type have higher specific activity than females which are XXY/+/*dsx*, interesexes of the XY/*dsx*/*dsx* or males, XYY/+/*dsx*. This effect is also more pronounced at the lower growth temperature. The basis for these effects is not clear at the present time, but it seems apparent that the levels of G6PD activity may be affected by several regulatory elements, some of which may not be X-linked. KOMMA (1968) has recently postulated another gene, M^{Zw}, which has the properties of changing the G6PD molecule such that its electrophoretic mobility is altered. This locus is linked to the *Zw* locus, but is apparently not allelic to it. The basis of the alteration effected by M^{Zw} is not understood, but it can be produced by mixing extracts from appropriate genotypes *in vitro*.

KOMMA (1968b) has given some evidence that the G6PD molecules from males and females of the same strain have different properties. These differences, which are observed in assays using crude extracts, include different electrophoretic mobilities, different thermal stabilitities, differences in stability at low EDTA concentration and differences in K_m for G6PD. It is somewhat difficult to interpret these differences, since they are noted in crude homogenates. It would be most interesting if sex differences in molecular properties of G6PD could be observed in purified enzyme preparations. This could mean that different closely linked genes function in males and females. Or it could mean that dosage compensation involves specific epigenetic modifications of the compensated enzyme. Suggestions of differences between sexes have also been noted by STEELE, *et al.* (1969), who observed that males which have the B-allele of G6PD have significantly higher specific activities than males with the A-allele. However, females having the B-allele or A-allele show little or no differences in G6PD activity. These authors also studied the effects of outcrossing on G6PD activity. F1 males obtained by crossing two different inbred strains show a significant increase, of about 30%, in G6PD level over either parent. Females in this Fl also show increases in G6PD levels, but of a smaller amount, approximately 15%. The increase in activity obtained upon outcrossing, and the differences between males and females, is not dependent on which allele is carried by the parents, nor is it dependent on the strains used. These authors have suggested that the activity increase seen on outcrossing is due to autosomal regulatory factors, and have interpreted the F1 male-female difference as indicating a differential response of the male genome to such factors.

Developmental Biology of G6PD

The gross distribution of G6PD in the various parts of the body has been examined by STEELE *et al.* (1969). The combined head-thorax region of either sex has about a 2.5-fold lower specific activity than the male abdomen. Female abdomens have about half the specific activity of the male abdomen. This latter difference is accounted for by the eggs in female abdomens. These make a large contribution to mass and protein content, but are relatively deficient in G6PD, thereby decreasing the specific enzyme activity of the female abdomen. This is also the most likely cause of the differences often observed between males and females when whole animals are assayed. Third instar larvae do not show this male-female difference.

WRIGHT and SHAW (1970) have determined the developmental history of G6PD during embryogenesis, and have observed relatively low, constant levels up until hatching. At this time the specific activity rises quite abruptly. STEELE *et al.* have measured the G6PD levels in eggs, larvae and pupae. They note a small increase from egg to combined first-second instar. Activity increases also during third instar. Young and old pupae are reported to have the same levels. This level is about half that reported for the adult.

6-Phosphogluconate Dehydrogenase (EC 1.1.1.44)

Genetics of 6PGD

YOUNG (1966) has examined seventeen strains of *Drosophila* for the electrophoretic mobility of 6-phosphogluconate dehydrogenase (6PGD). Twelve strains have a fast migrating form, 6PGD-A. Three strains have a slower migrating form, 6PGD-B. Two strains were found to be polymorphic, and showed both the A- and B-form.

Single males from a polymorphic strain have only one band of activity which is either in the A position or the B position. Females of these strains have either the A- or B-form, or show a three-band zymogram of 6PGD consisting of A, B and a band intermediate between A and B. Crosses between individuals from strains having the A form with individuals from strains having the B form yield male progeny that always inherit the 6PGD from their mother. Therefore, 6PGD like G6PD is produced by a sex-linked gene.

YOUNG (1966) has mapped the genes responsible for the electrophoretic variation of 6PGD to position 0.9± on the X-chromosome. This position is referred to as the *Pgd* locus. 6PGD has also been studied with respect to dosage compensation. The levels of activity in males and females are similar (KOMMA 1966; SEECOF *et al.*, 1969), therefore dosage compensation is operating. This enzyme does not respond to the mutant genes *tra* and *dsx* in a manner similar to G6PD. Flies homozygous for either of these genes have unaltered levels of 6PGD (KOMMA, 1966). 6PGD also does not show the outbreeding effect observed with G6PD (STEELE *et al.*, 1969). The presence of the intermediate band, presumably a heteropolymer of A and B, in heterozygous females demonstrates that the mechanism of dosage compensation in *Drosophila* is not of the X-inactivation type seen in mammals. In this case both alleles appear to be equally active in one cell, provided that there is no intercellular transport of subunits, which seems unlikely but not absolutely ruled out (KAZAZIAN *et al.*, 1965). SEECOF *et al.* (1968) have created flies with abnormal doses of the *Pgd* locus, and have studied the effects of these genotypes on enzyme activity levels. In these experiments, standard diploid females, standard hemizygous males, females with one *Pgd* locus translocated to chromosome three, females carrying a deletion close to but not covering *Pgd* and triploid females all show the same level of 6PGD activity. In all these cases, the ratio of autosomes to *Pgd* doses is what is expected for each sex. When this ratio is altered, changes in 6PGD levels are observed. Females carrying two normal X-chromosomes and a third dose of *Pgd* as an insertion on chromosome three have 1.33 times the standard level of activity. Males with one normal X, and the same insertion on the third chromosome, have 1.28 times the expected level. Females heterozygous for a deletion covering the *Pgd* locus have 0.61 times the level of standard activity. These results are similar in principle to those obtained by SEECOF *et al.* concerning G6PD levels.

Biochemistry of 6PGD

KAZAZIAN *et al.* (1965) have partially purified 6GPD by successive fractionations using ammonium sulfate, CM cellulose ion exchange chromatography and DEAE cellulose ion exchange chromatography. Studies with this partially purified enzyme have been conducted to determine the nature of the 3 isozyme bands, and the subunit relation in 6PGD. These authors have treated a mixture of partially purified form A and partially purified form B with propanedithiol, a disulfide bond reducing agent. After removal of propanedithiol, subsequent electrophoretic analysis showed the formation of the intermediate band usually seen in flies heterozygous at the 6PGD locus. Treatment of B-form alone did not produce additional bands. Mixing A and B without propanedithiol treatment did not generate the intermediate band. Therefore, the simplest model for the structure of 6PGD would appear to be a dimer or multiple of a dimer. The three-banded zymograms from heterozygous individuals, or from dissociated-reassociated enzyme mixture, represent the three possible combinations of the A and B subunits, i.e. AA, AB, and BB. KAZAZIAN (1966) has shown that multiple 6PGD are not size isomers. He subjected an extract which contained all three 6PGD bands to gel filtration chromatography on Sephadex G200. 6PGD activity chromatographs as a single symmetrical peak with an estimated molecular weight of 79000 ± 8000. Electrophoretic analysis of samples from different regions of the activity peak always showed the same amount of relative staining in each of the three isozyme bands, indicating no size-differences between them.

Alpha Glycerophosphate Dehydrogenases (EC 1.1.1.8), (EC 1.1.99.5)

Two enzyme systems which catalyse the interconversion of glycerophosphate and dihydroxyacetone phosphate have been described in *Drosophila*. These enzymes have recently been defined and distinguished by O'BRIEN and MCINTYRE (1972a, b). One system is glycerophosphate dehydrogenase (αGPDH), which is in the soluble fraction and is NAD-linked. The other, α-glycerophosphate oxidase (αGPO), is found in the mitochondria, and is NAD independent.

Biochemistry of αGPDH

GRELL (1967) initially observed that αGPDH exists as three isozymes. WRIGHT and SHAW (1969) observed only one band in extracts of larvae. This is the slowest migrating isozyme, αGPDH-3. Three bands of activity were observed in adults. These isozymes may represent enzyme with different physiological roles, as will be discussed below. The αGPDH-1 and αGPDH-3 can be distinguished by their different pH optima, as well as different electrophoretic mobilities. WRIGHT and SHAW (1969) eluted these two isozymes from starch gels after electrophoresis, and measured the pH optima of each in the direction of dihydroxyacetone phosphate reduction. The pH optimum of αGPDH-1 is 6.7–7.0, while αGPDH-3 has its optimum at pH 7.6. However, the isozymes are closely related because there is strong genetic evidence, to be discussed later, indicating that αGPDH-1 and αGPDH-3 have at least one common subunit.

O'BRIEN and MCINTYRE (1972a) have studied several of the biochemical properties of αGPDH. It shows a broad pH optimum curve with a maximum of 9.6, when the reaction is in the direction of glycerophosphate oxidation. The enzyme can be precipitated in the 45–60% cut of saturated ammonium sulfate. The molecular weight of αGPDH as determined by sedimentation in sucrose gradients was found to be 81000.

Biochemistry of αGPO

This enzyme can be distinguished from αGPDH by several criteria. In addition to its NAD independence and mitochondrial localization, it shows a bimodal pH optimum curve with maxima at 6.2 and 7.2. When solubilized from mitochondria and subjected to ammonium sulfate fractionation, it behaves in a highly lipophilic manner. The activity is found in a floating, fatty fraction in 30% saturated ammonium sulfate. Solubilized αGPO floats when subjected to sedimentation in sucrose gradients, and cannot be easily electrophoresed further, indicating its lipophilic nature. These properties are quite different from αGPDH. In addition, mutants which lack αGPDH have normal levels of αGPO.

αGPO appears to exist in two forms, as judged by the pH optimum curve and the observation that two zones of activity are observed by using isoelectric focusing in acrylamide gels. There is activity in the pH 7.2 and pH 6.6 fractions. These two forms are referred to as αGPO-1 and αGPO-2 by O'BRIEN and MCINTYRE (1972a). The existence of the two forms of αGPO may be of some physiological significance, since they show different developmental regulation (see below).

Genetics of αGPDH

GRELL (1967) found a variant of αGPDH with altered electrophoretic mobility. He has mapped the gene for this variation to 2–17.8. Analysis of a deficiency for the αGpdh locus shows it to be located between 25El and 26Cl on the cytogenetic map. O'BRIEN and MACINTYRE (1972b) have also mapped the αGpdh gene using the electrophoretic variant, and have found a map position of 2–20. The authors feel that this value is compatible with GRELL'S earlier position, because of errors associated in both measurements. O'BRIEN and MACINTYRE (1972b) have used the deficiency which covers the *αGpdh-1* locus to facilitate the isolation of chemically induced mutants. The rationale is similar to that used by GRELL *et al.* (1968) for isolation of ADH mutants. Five new alleles at the *αGdph-1* locus were obtained. One of these was a new electrophoretic variant. The other four have reduced (0–12% of wild-type) levels of αGPDH. Heterozygotes for the wild allele and either of three αGPDH-deficient alleles show dosage dependency, i.e. they have about 50% of the activity of wild-type flies. This is further evidence that the *αGpdh-1* locus is the structural gene for αGPDH, since dosage dependency appears to be a property of structural genes. One of the induced mutant alleles which result in αGPDH deficiency shows complementation with other alleles at this locus. When heterozygous with another deficient allele, more than additive enzyme levels are obtained. When in combination with a wild-type allele, essentially wild-type enzyme levels are seen. The interpretation given by O'BRIEN and MACINTYRE, of this inter-allelic complementation is that this allele produces an altered enzyme molecule that cannot make normal subunit associations with itself. However, in the presence of monomers that can associate properly, it can associate with them and contribute to the catalytic activity.

The analysis of isozyme patterns shown by electrophoretic variants and their hybrids has led to a discussion of the structure of αGPDH (WRIGHT and SHAW,

1969). Variants of αGPDH show a shift in the entire zymogram pattern i.e. αGPDH-1, αGPDH-2, and αGPDH-3 are affected. This means that all of the isozymes must have at least one subunit which is a product of the *αGpdh-1* locus. Two hypotheses of αGPDH structure can account for these results. The enzyme could be a homopolymer, a dimer (or multiple of a dimer) with the differences between the isozymic forms the result of an epigenetic modification of some of the enzyme molecules. The alternative hypothesis is that the enzyme is made up of two different kinds of subunits. One of these is common to each isozyme and is the product of the *αGpdh-1* locus. The differences in mobility of αGPDH-1, 2 and 3 would be the result of a difference in the nature of the non-shared subunit. Using the letter C to represent the common subunit and A and B the differing subunits, the isozymes could be represented as follows:

αGPDH-1 $(AA)_n(C)_n$,
αGPDH-2 $(AB)_n(C)_n$,
αGPDH-3 $(BB)_n(C)_n$.

Further investigations to distinguish between these hypotheses are needed and would be interesting, since αGPDH shows interesting developmental regulation and tissue distribution that could be more fully understood with a more complete knowledge of the structure of the isozymes. Purified αGPDH could be directly analyzed for the presence of more than one kind of subunit. It may be of interest to note that the second hypothesis of αGPDH structure predicts that electrophoretic variants of the second subunit could exist. Since the already fairly extensive work on the genetics and zymogram patterns of αGPDH has failed to uncover these, epigenetic modification of the enzyme may at this time be a more likely hypothesis.

Genetics of αGPO

As yet no information as to the gene(s) related to αGPO is available. However, one important conclusion has been reached by O'BRIEN and MACINTYRE (1972b). They have analyzed αGPDH deficient mutant with respect to αGPO activity. There is no significant effect of these mutations on αGPO activity, therefore establishing that the *αGpdh-1* locus is not a structural gene for αGPO.

Developmental Biology of αGPDH

The activity of αGPDH during development has been studied by WRIGHT and SHAW (1969, 1970), RECHSTEINER (1970b) and O'BRIEN and MACINTYRE (1972a). These investigators have all observed a similar developmental pattern, and have expressed their results in terms of activity per organism. Their data are summarized in Fig. 24. During embryogenesis, activity remains constant. A few hours before hatching, the levels of activity begin to rise. This rise coincides with the first appearance of the paternal form in embryos heterozygous for the fast and slow *αGpdh-1* alleles (WRIGHT and SHAW, 1969). Activity rises throughout the larval stages in a manner that correlates with growth. At pupation the rise in activity

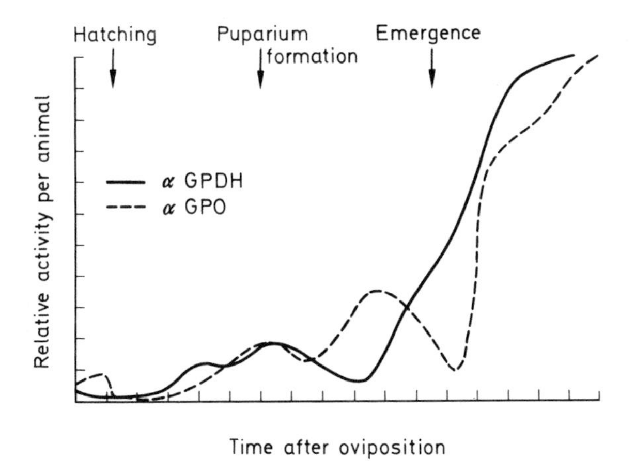

Fig. 24. Developmental profiles of α-glycerophosphate dehydrogenase and α-glycerophosphate oxidase. (α-GPDH redrawn from WRIGHT and SHAW, 1969; RECHSTEINER, 1970; O'BRIEN and MACINTYRE, 1972a. α-GPO redrawn from O'BRIEN and MACINTYRE, 1972a)

ceases, and the levels of αGPDH decrease for the first three days of pupal life. A day or so before emergence, αGPDH rises again. The rise in activity continues for a few days after emergence, until a maximum in the amount of αGPDH is reached in adult flies that is at five times that seen in larvae. WRIGHT and SHAW (1969) have noted that the rise in activity preceding emergence coincides with appearance of the major adult isozyme, αGPDH-1. As judged by the intensity of enzyme bands in gels, αGPDH-3 does not appear to increase at this time. This change in relative amounts of the isozyme is also reflected in change of the ratio of activity measured at pH 6.7, to that at 7.6. This ratio is about 0.2 in larvae, but about 1.2 in adults. As mentioned above, the larval isozyme, αGPDH-3, has a pH optimum of 7.6 and the predominant adult isozyme, αGPDH-1, has a pH optimum 6.7.

RECHSTEINER (1970b) has determined the tissue distribution pattern of αGPDH in larvae and in adults. Histochemical procedures were applied to frozen sections, and have demonstrated the presence of αGPDH in larval muscle, brain, gut and especially fat body. Dissection experiments have confirmed the high levels of activity in the fat body. Approximately 50% of the total activity of third instar larvae is found in the fat body. In adults, 75% to 85% of the total activity is found in the thorax. This includes the flight muscles, which in insects are known to contain high levels of αGPDH. The developmental history and tissue distribution studies suggest that the isozymes of αGPDH may play different metabolic roles. αGPDH-1 may be an enzyme predominantly in muscle and involved in NAD^+ regeneration and glycolysis. αGPDH-3 may be predominantly involved in lipid biosynthesis. The evidence for this functional differentiation of isozymes, while not as yet conclusive, can be summarized as follows. Larvae have mostly αGPDH-3, and most of this is in the fat body. Adults have mostly αGPDH-1, and most of this is in the thorax, presumably in the flight muscles. The increase in αGPDH-1 occurs around the time of muscle differentiation, and flies mutationally deficient for αGPDH have restriction in their ability to fly (O'BRIEN and MACINTYRE, 1972a). The αGPDH-3 that is found in adults is found primarily in the abdomen, where it might be involved in lipid biosynthesis (WRIGHT and SHAW, 1969). This hypothesis is further suggested by differences observed between males and females. RECHSTEINER (1970b) has observed higher αGPDH levels in females than in males. The zymograms of WRIGHT and SHAW (1969) indicate that αGPDH-1 band from males and females stain with the same intensity, but that the αGPDH-3 band of the females stain more intensely than that in males. This male-female difference could represent additional αGPDH for the synthesis of the lipoidal components of the eggs in the female abdomen. A more extensive discussion of the αGPDH functions and their interrelations with αGPO can be found in O'BRIEN and MACINTYRE (1972a). Reviews of the extensive research done on the α-glycerophosphate metabolism cycle in insects can be found in SACKTOR (1965 and 1970).

Developmental Biology of αGPO

O'BRIEN and MACINTYRE (1972a) have also studied the changes in αGPO activity per animal during development (Fig. 24). Activity rises throughout larval life and, following a dip around the time of pupation, rises again during pupal life. Following an abrupt fall just preceding emergence, a sharp rise in activity occurs in newly emerged adults. Changes in the relative amounts of αGPO-1 and αGPO-2 also occur during development. αGPO-2 is the major form found in larvae. It is the minor form found in adults. However, in adults, it is found in the head, thorax and abdomen. αGPO-1 on the other hand is minor, if present at all, in larvae. It finally becomes predominant in late pupae. In pupae and in adults it is only found in the thorax. These observations are consistent with αGPO-1 being the mitochondrial form of the enzyme involved with energy production in the flight muscles.

Lactate Dehydrogenase (EC 1.1.1.27)

The enzyme lactate dehydrogenase (LDH) has been extensively studied in vertebrates and invertebrates, but has only recently been described in *Drosophila* by RECHSTEINER (1970a). This enzyme was reported to exist in only one electrophoretic form in *Drosophila,* which is in sharp contrast to the vertebrate systems. In four different buffer systems which were employed by RECHSTEINER, only one band of activity could be observed. No difference was observed between larval LDH and adult LDH with respect to electrophoretic mobility. More recently PAPPAS and RODERICK (1971) and PAPPAS *et al.* (1971) have reported the existence of multiple bands of LDH activity in acrylamide gels. There is variation in enzyme band pattern during development. Thirty-seven strains of *D. melanogaster* have been screened with the hope of finding electrophoretic variants, but none have been found, and consequently there is as yet no information available on the genetics of LDH in *Drosophila* (RECHSTEINER, 1970a).

Biochemistry of LDH

LDH has been partially purified by successive protamine sulfate fractionation, batch DEAE cellulose adsorption and elution, ammonium sulfate fractionation and hydroxylapatite column chromatography (RECHSTEINER, 1970a). The LDH obtained is approximately 130-fold pure, and represents a recovery in the range of 20%. RECHSTEINER has used this purified enzyme to study some of the physical and enzymatic properties of LDH. The Stoke's radius for *Drosophila* LDH was determined by Sephadex G150 gel filtration and found to be 49 Å. The sedimentation coefficient was determined by sucrose density gradient centrifugation, and found to be 7.4 S. The molecular weight is estimated to be 150000. *Drosophila* LDH is specific for L-lactate, and shows approximately 6-fold higher activity with respect to NADH as compared to NADPH. The reaction in the reverse direction shows an approximately 100-fold faster rate using NAD^+ as compared to $NADP^+$. The activity is only slightly inhibited by high concentrations of pyruvate. The apparent K_m for pyruvate is 1.77×10^{-4} M and for NADH, 4.15×10^{-5} M. The pH optimum of the reaction of pyruvate to lactate is 6.5. The pH optimum for the lactate to pyruvate is 7.4.

The only suggestion of heterogeneity of *Drosophila* LDH comes from heat denaturation studies. The kinetics of inactivation at 62° C depart from the expected first order reaction. This may indicate the existence of at least two forms of LDH, with different thermal stabilities. NADH added to LDH makes it somewhat more heat stable. However, denaturation kinetics in the presence of the co-enzyme also suggests two forms. The enzyme which remains after a 93% loss due to heat denaturation differs only slightly from crude LDH in respect to pyruvate inhibition. It appears that if there are multiple molecular forms of LDH in *Drosophila*, the difference between them are subtle, and may or may not reflect the presence of multiple genes for LDH.

Developmental Biology of LDH

RECHSTEINER (1970b) has also investigated the developmental properties of LDH from *Drosophila*. His data are summarized in Fig. 25. This enzyme exhibits increases in activity during embryogenesis which appear to depart from a pattern common to several other of the dehydrogenase enzymes during this early stage of development. LDH shows a 50-fold increase in specific activity beginning early in embryogenesis. At this time other enzymes αGPD, IDH, 6PGD, G6PD, ADH and MDH are at apparently constant levels (WRIGHT and SHAW, 1970). The other similar enzyme which is reported to show major changes during embryogenesis is β-hydroxybutarate dehydrogenase (RECHSTEINER, 1970b). After hatching, animals show an increase in LDH activity that reflects the growth of the organism through the larval stages. The maximum activity levels are attained in late third instar larvae. Activity levels decline during the pupal stage, and a low level is found in adults.

Using histochemical procedures on frozen sections RECHSTEINER (1970b) has found that LDH is predominantly localized in muscle tissue. It is also found in the brain, ventral ganglion, imaginal discs and epidermis. No enzyme is demonstrable

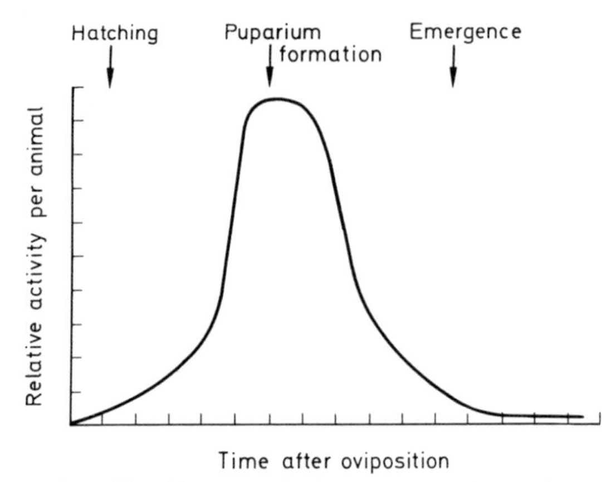

Fig. 25. Developmental profile of lactate dehydrogenase. (Redrawn from RECHSTEINER, 1970)

in the hind gut, midgut, fat body or Malpighian tubules. This distribution of enzyme has been confirmed using dissected organ histochemistry, except that isolated epidermis did not stain. This difference could reflect permeability barriers in epidermal cells to one or more of the reagents used in the whole organ technique. RECHSTEINER has suggested that the developmental fluctuations and tissue distribution of LDH may indicate its possible role in NAD^+ regeneration during periods of relative anerobiosis. In the adult muscle this role is likely to be performed by the high levels of αGPDH (see above).

Isocitrate Dehydrogenase and Aconitase (EC 1.1.1.42), (EC 4.2.1.3)

FOX (1971) has recently studied the properties of the NADP-linked isocitrate dehydrogenase (IDH) from *Drosophila*. In other organisms two enzymes having IDH activity are commonly found. One is NADP-linked and is found in the soluble fraction after differential centrifugation, the other IDH is an NAD-linked form associated with mitochondria. At the time of writing only the soluble IDH has been observed in Drosophila. Attempts have been made to study a mitochondrial IDH, but they have proven unsuccessful (FOX, 1971). The soluble, NADP-linked IDH has a pH optimum in TRIS-HCl buffers at pH 8.5. It is specific for NADP and has a requirement for divalent cations, either Mg^{++} or Mn^{++} are sufficient.

Genetics of IDH

FOX (1971) has discovered variants of IDH with altered electrophoretic mobility. A cross between flies having the "fast" form of IDH and flies having the "slow" form yields heterozygous progeny having three bands of activity. These are the two parental bands, and a darker staining band with intermediate mobility. The gene responsible for this variation has been mapped to 3-27.1 ± 0.4.

Developmental Biology of IDH

IDH shows an ontogenetic pattern typical for several dehydrogenases (Fig. 26). Activity is fairly low in eggs, and remains constant through most of embryogenesis. Around the time of hatching, total activity per organism begins to rise.

The first evidence of paternal gene function is also noted at this time, as judged by the appearance of the paternal enzyme band in animals heterozygous at the *Idh* locus (WRIGHT and SHAW, 1970). Activity gradually rises during larval life. This rise is approximately correlated with growth. Maximum total activity is found at about 96 hrs after oviposition, while the peak in specific activity is at 82 hrs. IDH

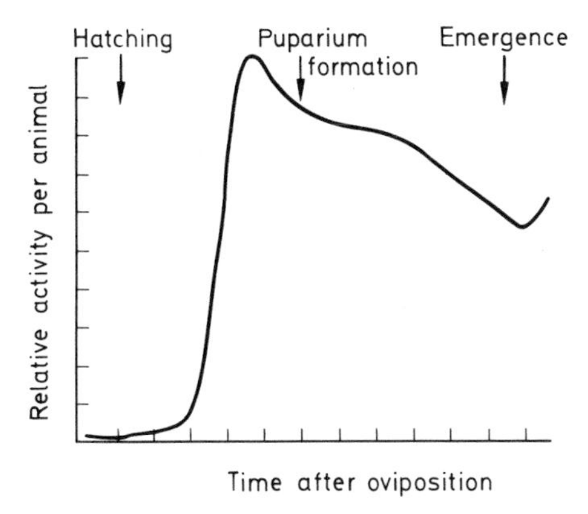

Fig. 26. Developmental profile of NADP isocitrate dehydrogenase. (Redrawn from FOX, 1971)

activity declines gradually through most of the pupal stage, and rises again to a second maximum in emerged adults (FOX, 1971). STEELE *et al.* (1969) have observed that adult males have a specific IDH activity 30% higher than females. This is most probably due to the presence of eggs with low amounts of IDH in the female abdomen.

FOX *et al.* (1972) have studied the tissue distribution of IDH and aconitase, the enzyme which precedes IDH in the citric acid cycle. In larvae, these enzymes are found in the fat body, carcass, combined intestine and Malpighian tubules and combined brains, salivary gland and imaginal discs. In adults, activity of both is found in the head, thorax and abdomen. Over 68% of adult IDH is in the abdomen, while 79% of adult aconitase is in the thorax. Developmental changes and tissue differences were observed in the three-banded IDH isozyme pattern. The meaning of these differences is not clear at present. Aconitase shows a five-band isozyme pattern. Tissue differences in this pattern were not observed. These authors observed that approximately 34% of the IDH and 15% of the aconitase is found in the mitochondrial fraction obtained by differential centrifugation. The isozyme pattern of IDH from the mitochondrial and supernatant fraction is the same. The isozyme pattern of aconitase from the supernatant fraction shows the 5-band isozyme pattern. The mitochondrial fraction has three of these bands, but is missing the two most anodally migrating aconitase isozymes.

Malate Dehydrogenase (EC 1.1.1.37)

Two enzymes with malate dehydrogenase (MDH) activity have been described in *Drosophila*. One is found in the supernatant fraction after centrifugation, and is called soluble MDH (s-MDH). The other is a mitochondrial associated enzyme (m-MDH).

Biochemistry of MDH

Both forms have been purified and partially characterized with respect to their enzymatic and physical properties by McReynolds and Kitto (1970). Purification has been effected by ammonium sulfate fractionation, followed by gel filtration on Sephadex G-100. Ion exchange chromatography on CM cellulose is then used to separate the m-MDH and s-MDH, and the two enzymes are further purified. Preparations of purified m-MDH show one protein band on cellulose acetate electrophoresis, and sediments as a single peak in the ultra-centrifuge. Purified m-MDH shows catalytic activity comparable to MDH purified from other systems. Purification is somewhat over 400-fold with a yield representing 3% of the total homogenate MDH activity. The sedimentation coefficient of m-MDH determined by analytical ultracentrifugation is 4.03. Preparations of s-MDH have been obtained that are about 90% pure as judged by the appearance of several protein contaminants being evident in cellulose acetate electrophoretic gels. The molecular weight of both m-MDH and s-MDH were determined to be about 68000 by gel filtration on Sephadex G-100. The two forms of MDH can be distinguished by immunochemical techniques, kinetic properties and electrophoretic mobilities upon starch gel electrophoresis. Antibodies were prepared against m-MDH from *D. virilis*. This antibody inhibits m-MDH from *D. virilis* and to a degree, the m-MDH from *D. melanogaster*. However, this antibody has no effect against s-MDH from *D. virilis*. Anti m-MDH reacts with m-MDH, using double diffusion in agar plates, but shows no precipitin band against s-MDH. The m-MDH is more susceptible to substrate inhibition by high concentrations of oxaloacetate than is s-MDH. The K_m for malate is 2×10^{-2} M using m-MDH, and 8×10^{-3} M for s-MDH. The K_m for oxaloacetate for m-MDH and s-MDH are similar, being approximately 4×10^{-5} M. The two forms of MDH can also be differentiated with respect to their utilization of the co-factor analogues acetylpyridine adenine dinucleotide and acetylpyridine hypoxanthine dinucleotide. s-MDH is about twice as active as m-MDH with acetylpyridine hypoxanthine dinucleotide, when compared to their rates using NAD^+ as a co-factor. However, m-MDH is relatively much more active with acetylpyridine adenine dinucleotide than is s-MDH. The pH optimum of both forms for oxaloacetate reduction is 8.5. The pH optimum for malate oxidation is 9.5 using m-MDH, and 9.0 using s-MDH. The two forms show differences in stability at 57° C, s-MDH being somewhat more stable. The two forms have differences in electrophoretic mobility on starch-gels at pH 7.0. Therefore, the identification and separation of the two MDH enzyme systems seems clear. With this supporting biochemical information it should be possible to get answers to questions concerning the genetics of the mitochondrial form and the genetic relation between the two forms, if any.

Genetics of MDH

Electrophoretic variants of the soluble form of MDH from *D. melanogaster* have been found by several investigators. O'BRIEN and MACINTYRE (1969) have reported a map position of 2-35.3. Map positions of 2-40.1 and 2-40.2 have been reported by ANDERSON and GRELL respectively (quoted in FOX, 1971). Hybrids between flies having the two alternative electrophoretic forms show a typical three-band pattern having both the parental bands and a stronger staining band of intermediate mobility. HUBBY and NARISE (1967) have reported the *in vitro* formation of the hybrid pattern by freezing and thawing mixtures of the two parental types in 1.0 M NaCl.

Glutamate Dehydrogenase (EC 1.4.1.2)

Glutamate dehydrogenase (GDH) catalyzes the interconversion of glutamate and ketoglutarate, and hence is an important link between amino acid metabolism and carbohydrate metabolism. GDH from *Drosophila* third instar larvae has been studied with respect to its enzymatic properties by BOND and SANG (1968). The enzyme is associated with the mitochondrial fraction obtained by differential centrifugation. It can be partially removed by sonication, 1% digitonin treatment or washing with buffer.

BOND and SANG (1968) have studied several aspects of the GDH catalyzed reaction using GDH, which was solubilized by washing mitochondria with buffer. The rate of the reaction in both directions, i.e. from glutamate to ketoglutarate and the reverse, are much higher using NAD^+ and NADH respectively than with NADP or NADPH. Kinetic analyses have demonstrated effects on the reaction by various purine nucleotides. AMP, cyclic 3′–5′ AMP, cyclic 2′–3′ AMP and ADP activate the enzyme. GTP, GDP, and ATP inhibit the enzyme to varying degrees. As with GDH from other sources *Drosophila* GDH is inhibited by diethylstilbestrol and zinc. BOND and SANG have interpreted their kinetic analyses as indicating that the primary function of GDH in *Drosophila* is that of energy production from amino acids. PAPPAS *et al.* (1971) have reported that isozymes of this enzyme can be observed in acrylamide gels. The basis of these multiple bands was not further investigated, and their nature is unclear at present. There is as yet no available information on the developmental or genetic aspects of this enzyme.

β-L-Hydroxyacid Dehydrogenase

BORACK and SOFER (1971) have characterized an enzyme from adult flies which catalyzes the oxidation of 1-β-hydroxyacids. This enzyme uses the 1 isomer of β-hydroxybutyric acid as a substrate, and is even more active with the 1 isomer of gluconic acid. It is similar in some respects to the pig kidney enzyme, and on this basis might be expected to oxidize other l-β-hydroxyacids as well. The enzyme from *Drosophila* uses NAD as a coenzyme, and has an apparent K_m of 0.25 mM NAD. BORACK and SOFER have purified the enzyme 2600-fold, with a yield of 50%. The preparation obtained shows a single protein band when subjected to acrylamide gel electrophoresis. Molecular weight estimations were performed by sedimentation in sucrose gradients. The determined S value was 4.4, which corresponds to a M.W. of 6.3×10^4.

There is no information yet available related to the genetics of this enzyme. Electrophoretic variants have not been reported. The physiological role of this enzyme also remains to be determined.

Enzymes of Tyrosine Metabolism

To date three enzyme activities in *Drosophila* have been described which function in tyrosine metabolism. These are phenol oxidase, dopa decarboxylase and dopamine-N-acetylase. These enzymes are relevant to several different functions, including melanization or black pigment formation, sclerotization and subsequent tanning, and possibly neural transmitter synthesis. These steps in the pathways are shown in Fig. 27. Melanin is formed by the oxidation of tyrosine to a quinone precursor of the indole subunits which make up melanin. Tyrosine is also a precursor for the as yet unidentified compound(s) which are involved in the hardening and darkening of the cuticle, enabling the cuticle to serve as an effective exoskeleton. These steps, hardening and darkening, are closely related, and are referred to as sclerotization and tanning respectively. It is beyond our scope to consider in complete detail the extensive work that has been done on the structure and sclerotization of the cuticle. MITCHELL *et al.* (1971) have recently described the structural changes which the cuticle undergoes during molting. Aspects of the biochemical process involved in sclerotization have been amply reviewed by PRYOR (1962) and BRUNET (1967). Although the details of sclerotization have not been studied extensively in *Drosophila*, analogies to other insects indicate some general patterns which occur during this process that may be expected to apply to *Drosophila*. Sclerotization involves reactions between amino groups of a cuticular protein and orthoquinones, which are produced by the action of an oxidase using dopa or some other metabolic derivative of tyrosine as a substrate. The resultant cross-linked quinone-protein structure is very rigid, and gives the insect cuticle its strength. The nature of the specific reacting protein is obscure, and details of its structure and function are unclear. The information available has been summarized by BRUNET (1967). The reactant protein amino groups could be from N-terminal amino acids, or the epsilon amino groups of lysine.

Quinone derivatives of a variety of compounds have been assigned the role of tanning agents in various insects. These include 3,4-dihydroxybenzoic acid, 3,4-dihydroxyphenylacetic acid and N-acetyl dopamine, among others. Reviews of tanning agents and their occurrence have been given by BRUNET (1963), and KARLSON and SEKERIS (1964). One common feature of compounds thought to be tanning agents is the lack of a functional amino group on the alpha carbon atom. It is apparently necessary that the tanning agent have the alpha amino group blocked or removed so that upon oxidation to a quinone, indole formation and subsequent melanin production will not occur.

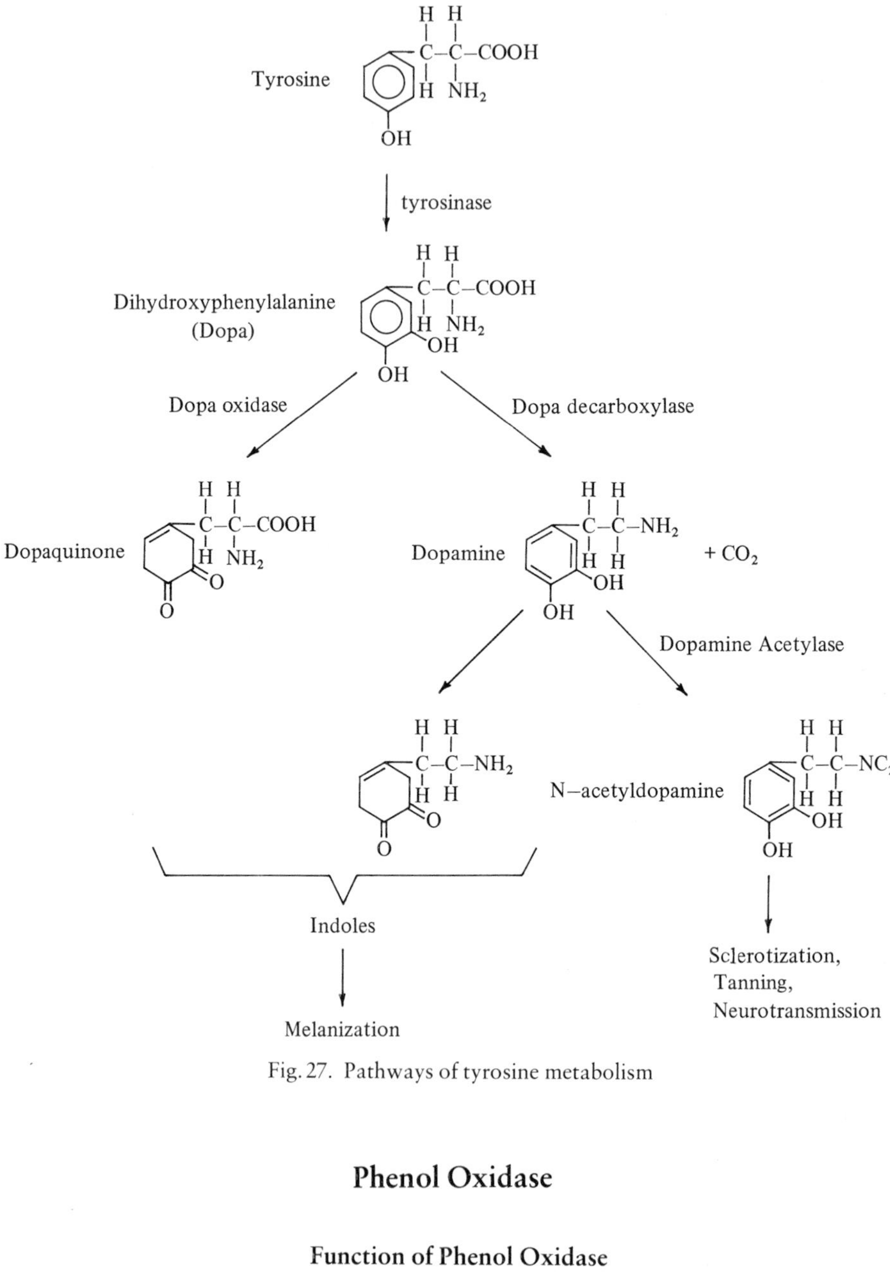

Tyrosine

H H | C–C–COOH | H NH_2 OH

tyrosinase

Dihydroxyphenylalanine (Dopa)

H H | C–C–COOH | H NH_2 OH OH

Dopa oxidase

Dopa decarboxylase

Dopaquinone

H H | C–C–COOH | H NH_2 O O

Dopamine

H H | C–C–NH_2 | H H OH OH

+ CO_2

Dopamine Acetylase

H H | C–C–NH_2 | H H O O

N–acetyldopamine

H H | C–C–NC_2H_5 | H H OH OH

Indoles

Melanization

Sclerotization,
Tanning,
Neurotransmission

Fig. 27. Pathways of tyrosine metabolism

Phenol Oxidase

Function of Phenol Oxidase

The generic name phenol oxidase is used here to include enzymes which use monohydroxyphenols, e.g. tyrosine, dihydroxyphenols, e.g. dopa or both as substrates. The names tyrosinase or dopa oxidase have also been used to describe these individual activities. Tyrosinase was one of the first enzyme systems to be studied in *Drosophila*. GRAUBARD (1933) took the then new knowledge regarding tyrosinase function in melanin formation, and attempted to apply it to the many body color variants of *Drosophila*. His aim was to understand the physiology of these body color genes. This analysis, however, gave no indication of any clear relation between the activity of tyrosinase and any of the mutants that were tested. The lack of relation between phenol oxidase activity, as measured in crude extracts, and mutations which seem to affect phenol oxidase activity *in vivo*, i.e. body color mutants, is not as surprising as it might seem. The expected relation between phenol oxidase activity and genes affecting body color depends on the assumption that the *in vivo* function of phenol oxidase is melanin production. While probably valid, this assumption obscures what are probably far more important functions of these enzymes, namely cuticular tanning. Melanin production is most likely of secondary physiological consequence when compared to the tanning of the cuticle. The separation of these functions was noted some years ago by WADDINGTON (1941), who compared the development of pigmentation in several mutants and pointed out that the pigmentation process consisted of two temporally overlapping phases, a browning phase and a blackening phase. These corresponded with tanning and melanization respectively, and are presumably linked by means of a common precursor.

The two functions, melanization and sclerotization, which are attributed to phenol oxidase have led to the question whether there are separate enzymes for each of these functions, or whether one enzyme system performs both functions. This has yet to be clearly answered. BRUNET (1963) gives some of the arguments for the separate enzyme hypothesis. However, there are some indications from the work on *Drosophila* that the same enzymes(s) could be involved in both processes. Several phenol oxidases have been reported in *Drosophila* (see below). Only one activity, however, has been found that converts tyrosine to dopa. This conversion is most likely the first reaction leading to the tanning agent and to melanin synthesis, and an enzyme catalyzing this step would presumably function in both pathways. The dihydroxyphenoloxidases could be specific for one pathway or another. However, there are no data yet available to indicate this. The most attractive hypothesis at the present time predicts that the choice for melanization or sclerotization is not determined by the phenol oxidases. Instead, it would be controlled by the presence or absence of enzymes further along the pathway in the synthesis of the tanning agent.

Another interesting aspect of the regulation of phenol oxidase function is the suggestion by several scientists that there may be a special storage form(s) of the enzyme's substrate, tyrosine. MITCHELL and LUNAN (1964) have found tyrosine-0-phosphate in third instar larvae. The kinetics of its metabolism are consistent with the interpretation that tyrosine-0-phosphate serves as a storage form of tyrosine which is used in tanning (LUNAN and MITCHELL, 1969). RIZKI and RIZKI (1959) have observed rectangular crystal inclusions in certain hemocytes, which become labelled when radioactive tyrosine is fed to the larvae. These crystals can be observed to dissolve upon cellular damage, and could be substrates for phenol oxidase. HENDERSON and GLASSMAN (1969) have described a pigment producing factor, PPF, which leads to pigmentation of pupae when injected. This substance is described as being macromolecular and non-protein in nature, and is hypothesized to represent a storage form for tyrosinase substrates. The relation, if any, between these various possible storage forms of phenol oxidase substrates is not clear.

Biochemistry of Phenol Oxidase

The assay usually employed in the measurement of phenol oxidase activity consists of using dopa as a substrate and following the production of its oxidation product, dopaquinone, spectrophotometrically (HOROWITZ and FLING, 1955; MITCHELL, 1966). A consideration of the kinetics of this assay points out the first peculiarity of this enzyme. The rate of dopachrome production is linear for a short period of time. The duration of linearity is sufficient to provide a satisfactory assay, but is dependent on enzyme concentration often necessitating an appropriate dilution series. The reason for the plateau in dopachrome production is not exhaustion of substrate, but rather it appears as if the enzyme is used up during the reaction (MITCHELL, 1966). There are two possible reasons for this behavior. The enzyme may be oxidizing itself, that is, peptide tyrosines may be serving as substrates. Their oxidation could lead to denaturation of the enzyme. Secondly, the product of the reaction, dopaquinone, is highly reactive, and it may be "tanning" the enzyme molecule in a manner analogous with cuticular tanning. MITCHELL (1966) has suggested that these properties may be related to the biological function of the enzyme. The poor catalytic properties of phenol oxidase do place some limitations on the kinds of experiments that can be performed. More importantly, however, they should serve as a note of caution when trying to make quantitative statements about phenol oxidase acitivity using essentially qualitative procedures, e.g. measuring activities by observing activity bands in gels after electrophoresis.

One of the most striking features of phenol oxidase in *Drosophila* and other insects is the phenomenon of activation. It was observed by HOROWITZ and FLING (1955) and OHNISHI (1954) that phenol oxidase activity could not be detected in extracts of *Drosophila* immediately upon preparation. However, upon standing at 0° C, activity appeared in these extracts. HOROWITZ and FLING noted that the kinetics of this activation process followed a sigmoid curve, and suggested a modified auto-catalytic mechanism for activation. A series of recent investigations by MITCHELL and co-workers has served to point out the degree of complexity of the phenol oxidase system, and has provided some initial information needed to understand the activation phenomenon. MITCHELL and WEBER (1965) found that there were at least five components involved in the activation system. Three of these are referred to as the A components, A1, A2, A3. These can be electrophoresed in acrylamide gels and then activated *in situ* to yield phenol oxidase containing zones. The bands of activity can be demonstrated by incubation with the appropriate substrate, and observing the deposition of a melanin-like product. Al shows activity when either tyrosine or dopa is used as substrate, while A2 and A3 are specific for dopa, and show little or no activity towards tyrosine. The fraction used to activate the A components after electrophoresis is called P. This fraction is a suspension of a 0–41% ammonium sulfate precipitate of a pupal extract. Upon further investigation, this fraction has been found to contain two components, referred to as P and S. These function in the activation process, but have no phenol oxidase activity themselves. MITCHELL *et al.* (1967) have hypothesized that activation involves an assembly of subunits represented by the various components, into a finished molecule having phenol oxidase activity. The exact nature and function of each component has not been determined. Probably all are proteins, and partial purification has been attempted for the A and S components. Little is known about P, except that it is necessary for activation, is heat labile, and can be fractionated with ammonium sulfate. Representation of the activation process is shown in Fig. 28.

The function of each of the various components in phenol oxidase activation is not fully understood, but preliminary data are available which enable at least a partial explanation. Currently it appears that the A components serve in some way as pro-enzymes, as indicated by the existence of the gel assay and the demonstration of substrate specificities for them. The function of component S has been investigated, and it appears that S functions catalytically during activation. Using partially purified S, it can be shown that the rate of activation of phenol oxidase is dependent on S concentration, but that the final level of activity obtained is inde-

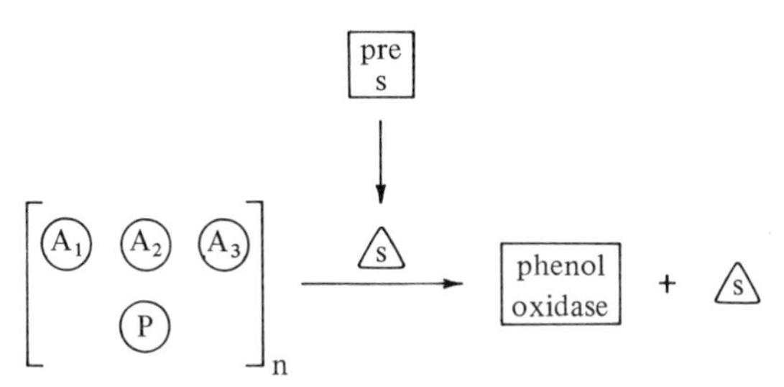

Fig. 28. Phenol oxidase activation

pendent of the amount of S present. It is also possible to recover S from an activation reaction, and show that there is no substantial loss of S activity during activation (SULLIVAN and MITCHELL, 1969). Another complicating aspect of the behavior of phenol oxidase is the observation made by GEIGER and MITCHELL (1966) that S occurs as an inactive precursor form, and undergoes an activation process of its own. The basis of this activation reaction is unknown at present. Phenol oxidase activation, then, can be summed up as a combination of two consecutive reactions. First, the precursor of S is activated. Active S then functions catalytically with the other components of the system to produce phenol oxidase activity. A consideration of these reactions indicates why an autocatalytic mechanism was at first suggested to explain phenol oxidase activation. Upon preparation of an extract, S activation is initiated, activated S begins to function, and phenol oxidase activity commences. As time proceeds, more active S is being produced, thereby increasing the rate of phenol oxidase activation. Phenol oxidase activation continues until the exhaustion of the other components, at which time the activation curve plateaus. This mechanism generates an activation curve approximating a sigmoid form similar to that produced by an autocatalytic mechanism. This mechanism may eventually be found to be somewhat oversimplified, but it seems at present to be the most satisfactory explanation of this complicated phenomenon. Conclusive answers await the purification of each component and *in vitro* study of the biochemical mechanisms involved in the function of each component.

Studies on the protein chemistry of the phenol oxidase molecule have been greatly hindered by its aggregation properties. Aggregation of phenol oxidase occurs coincident with enzyme activation. It has not been possible to separate these two processes, therefore the only enzyme available for study is in the form of quite large aggregates. The nature of this aggregated enzyme has been studied by MITCHELL *et al.* (1967). Sedimentation velocity experiments have shown that these enzyme aggregates are very large and very heterogenous in size. However, it should be noted that the size of a given aggregate is stable, as indicated by repetitive velocity sedimentation analysis. These aggregates can be viewed by light microscopy, and undergo a self-melanization when small amounts of dopa are added to them. The aggregates appear as amorphous masses composed of units which are about 1μ in diameter. Activated phenol oxidases have also been studied by equilibrium sedimentation in sucrose gradients. Seven components with discrete densities have been observed. The density of all of these components is somewhat less than would be expected for pure protein, indicating the presence of varying amounts of lipid material. The significance of these seven isozymes is not clear at present. However, the patterns of their relative distribution during development are highly specific and regular, indicating that the appearance of these isozymes could be related to different functions during development. Also, the pattern of enzyme distribution in equilibrium density gradients is different in extracts prepared from different body color mutants, suggesting that whatever the basis for the relative amounts of the seven isozymes is, it may be related to their *in vivo* function.

Another type of phenol oxidase has been described by YAMAZAKI (1969) in *Drosophila virilis*. This enzyme is found in the cuticle, and it is assayed by using a suspension of purified and ground cuticle obtained from white prepupae. This localization is evidence of it being involved in sclerotization and tanning of the cuticle. It can be differentiated from the phenol oxidase described above on the basis of several properties. It is more stable, and has a higher pH optimum, 7.5. It is not sensitive to carbon monoxide or thiourea. It does not use tyrosine as a substrate. It uses a variety of different compounds as substrates, including in order of decreasing utilization: 3-4 dihydroxyphenylacetic acid, p-aminophenol, pyrogallol, dimethyl-p-phenylenediamine, hydroquinone, catechol, dopamine, 3-hydroxykynurenine and p-phenylenediamine. Any structural or functional relation to other phenol oxidase systems is not evident at this time.

Genetics and Developmental Biology of Phenol Oxidase

It is common practice in developmental biochemistry to measure levels of enzymes in extracts prepared from organisms at different stages of development, and to use this information to draw conclusions about the relative activity of the genes which produce the enzymes. In recent years it has become more commonly appreciated that the measured level of enzyme activity is a function of both the rate of synthesis of the enzyme and its rate of degradation (see SCHIMKE and DOYLE, 1970), so that it is now recognized that changes in enzyme activity during development can reflect changes in the sum of activity of several genes: the structural genes for the enzyme, regulatory genes, or genes whose products control enzyme degradation. However, in the phenol oxidase system there are several lines of evidence that indicate that the maximal activity measured in extracts may only be indirectly related to the sum of the activities of the types of genes mentioned above. It should be recalled that no phenol oxidase activity can be measured in fresh extracts of *Drosophila*. This is true even at times during development when it is obvious that enzyme function is occurring *in vivo*. What can be measured is the potential for activity. Activity then is defined as the maximum activity obtained after allowing complete *in vitro* activation to occur. This implies that for some reason the enzyme phenol oxidase is not extractable, but that the components prior to activation are extractable. Clearly, in order to draw conclusions about the function of genes involved with phenol oxidase production, knowledge of both the amount of a component in the finished enzyme and the amount not yet activated is needed. No way is yet available to assess the former. There appears to be no compelling reason for assuming a constant ratio of component in enzyme to free component.

MITCHELL *et al.* (1967) have suggested that it is not possible to measure activated enzyme in fresh homogenates, because as phenol oxidase is formed it becomes fixed as part of an insoluble structure. The observations of self-inactivation, aggregation and self-melanization previously mentioned support this idea. It appears then, that measurements of the amount of enzyme potential which can be activated measures the amount of enzyme components "left over" from melanization or tanning, or components which have been made in anticipation of use in these processes at some future time of development, but which have not yet reached the site where they are activated and made insoluble. Most likely a sum of these is what is usually measured, and the relative contributions of the two are apt to vary during development. Further support for the fixing and subsequent loss of activated phenol oxidase through function comes from several observations, which show an inverse relationship between the apparent *in vivo* function and the amount of measurable enzyme activity obtained after activation *in vitro*. GRAUBARD (1933) noticed that the mutants *black (b)* and *ebony (e)* had less activity than did wild-type, whereas *yellow (y)* had higher activity. MITCHELL (1966) observed higher potential activity in extracts of the lightly pigmented mutant *straw (stw)*. MITCHELL *et al.* (1967) observed that *Blond (Bld)* and *yellow (y)* newly emerged flies and late pupae, which have pigment deficiencies, have more activity than wild-type, whereas *black* and *ebony*, dark body color mutants, have somewhat less than wild-type. YAMAZAKI and OHNISHI (1968) have observed that lines whose individuals contain melanized masses, the so-called melanotic tumors, have significantly less phenol oxidase activity *in vitro* than do genetically comparable lines without melanized masses. The most convincing evidence concerning loss of activity through function comes from the experiments of MITCHELL (1966), in which he induced phenocopies of the mutant *Blond (Bld)* by heat shock. The blond phenocopies have higher activities than non-treated wild-type controls during the time of pigment formation. The basis for this difference must be physiological, since the phenocopies are genetically identical to the controls. The higher activity probably represents an excess of enzyme remaining after the failure to pigment or tan normally.

Further evidence that the measured levels of activity have only secondary relation to the function of phenol oxidase *in vivo* comes from the data of LEWIS and LEWIS (1961), who measured the amount of enzyme activity in different strains of *Drosophila* and in many mutants. The activity levels in these strains were seen to vary over a range of close to forty-fold. The strains at the extremes of this range appear similar with respect to body pigmentation and cuticle.

It appears that the levels of phenol oxidase activity after maximum activation in extracts of *Drosophila* are related to several variables, in addition to the activity of genes responsible for the synthesis and degradation of the enzyme itself. Levels of activity *in vitro* are probably dependent on, and inversely related to, factors which control activity *in vivo*, such as amount or availability of substrate. Activity *in vivo* is also probably dependent on the incorporation of enzyme into the proper structural situation at the proper time. The mechanisms involved in these processes are undoubtedly complex, and are obscure at present.

These indications that measurement of phenol oxidase potential activity indicate some unknown and variable fraction of the actual amount of enzyme produced in the animal, place severe restrictions on the interpretation of quantitative differences in levels of activity which might be seen in making comparisons between various mutants and wild-type strains. Experiments recently performed on various alleles of the *lozenge* mutant *(lz)* serve to point out the caution that should be used in this regard. PEEPLES *et al.* (1968) measured the A1 and the combined A2–A3 components which could be activated in gels after electrophoresis of extracts from single animals. They observed an absence of the A1 band in the mutant *lozenge-glossy* *(*lz^g*)*. They were also unable to detect any tyrosine oxidation using the spectrophotometric assay. Subsequent investigations (PEEPLES *et al.* 1969a, b) have shown deficiencies for tyrosine and dopa oxidation in flies carrying other alleles of the *lozenge* series. These comparisons were performed by using the appearance of bands in polyacrylamide gels after electrophoresis, activation, and then incubation with substrate as an indication of activity. PEEPLES and co-workers have shown a correlation of phenotypic severity of the respective *lozenge* alleles, and lowering of the activity when judged by the visual appearance of stained gels. Several of the more severe *lozenge* alleles are characterized as having no tyrosinase activity at all. Four *lozenge* alleles are reported as having neither the ability to oxidize tyrosine nor dopa. These observations are puzzling. *Lozenge* mutants do not have major defects in body pigmentation, i.e. they have apparently normal melanin production *in vivo*, and therefore presumably have the enzyme in sufficient quantities for this reaction. The only enzyme responsible for metabolizing tyrosine to dopa that has been described in *Drosophila* involves the A1 derived phenol oxidase. An alternate pathway cannot be excluded, however, a mutant completely lacking A1 tyrosinase activity would be likely to be lethal, since it would not be able to form sclerotinized cuticle at all. Furthermore, it seems almost certain that mutants lacking both tyrosine and dopa oxidizing ability would be lethal. *Lozenge* mutants do have cuticle related structural abnormalities which are consistent with altered or local deficiencies in sclerotization. However, these do not seem to be of sufficient severity to expect the complete lack of enzyme. It appears that for some yet unknown reason the phenol oxidase A components may be difficult to extract, or more labile than normal. However, the proposal of PEEPLES *et al.* (1969b) that the *lozenge* locus is a structural gene for a polypeptide of an A component seems premature.

The most convincing evidence for the role of a specific gene's direct involvement in the production of phenol oxidase comes from the experiments of LEWIS and LEWIS (1963). They have described a locus on the second chromosome in a strain that has low phenol oxidase activity. This locus has been called *alpha*, and was mapped at 52.4 of the second chromosome. The phenol oxidase produced in strains homozygous for α has an increased thermolability. On the basis of this qualitative

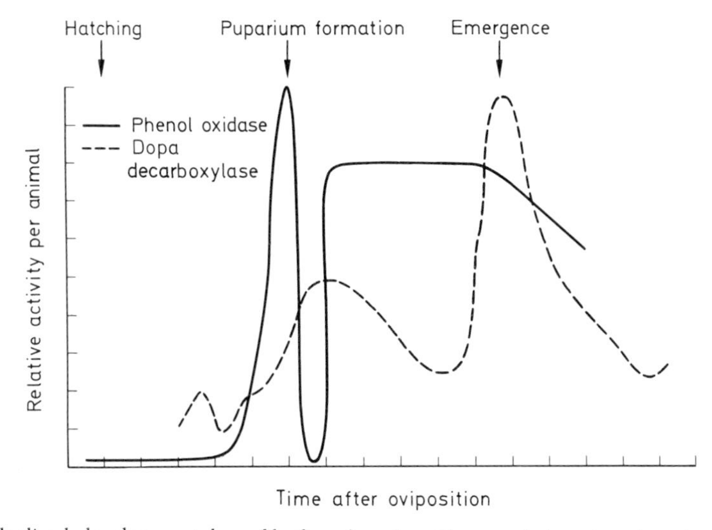

Fig. 29. Idealized developmental profile for phenol oxidase and dopadecarboxylase. Phenol oxidase, data redrawn from MITCHELL (1965) and GEIGER and MITCHELL (1966). Dopa decarboxylase, data redrawn from LUNAN and MITCHELL (1969) and MCCAMAN *et al.* (1972)

difference in the enzyme molecule, LEWIS and LEWIS tentatively concluded that the locus may be a structural gene for phenol oxidase. They have also identified several other genetic factors which appear to be responsible for controlling the levels of phenol oxidase activity. These factors are located on chromosomes two and three. Further characterization of these factors has not as yet been reported.

MITCHELL has pointed out that the body color mutants *ebony, black, yellow* and *Blond* can be considered mutations of genes which serve a regulatory function in the phenol oxidase system. The mutant alleles of these genes produce altered patterns of enzyme activity in sedimentation equilibria-centrifugation experiments, and produce obvious alterations in the *in vivo* function of the enzyme (MITCHELL *et al.* 1967). It seems that the use of the term "regulatory genes" to describe these phenomena is appropriate, as long as it is understood that no mechanism of function of the regulatory genes is implied. The genes are called regulatory in a broad sense, and analogy to the better understood regulatory genes in prokaryotes is not intended, and should not be assumed.

While measurements of phenol oxidase activities during development (Fig. 29) are difficult to interpret in a genetic sense, they have been useful in furthering our understanding of the function of this enzyme system. MITCHELL has found activity in animals as early as first instar larvae. Activity levels stay low but detectable throughout the first, second and first part of the third larval instar stages. At mid third instar the levels of activity begin to rise, and maximum levels are attained at about the time of puparium formation (GEIGER and MITCHELL, 1966). Activity remains approximately constant through pupal life, declining gradually as the animal approaches emergence. However, there is one brief period during the prepupal stage, 4 to 8 hrs after puparium formation, when no phenol oxidase is measurable (MITCHELL, 1966). This temporary deficiency is apparently due to a transient lack of active S component (GEIGER and MITCHELL, 1966). Following emergence there is a gradual decline in phenol oxidase activity. Flies 10 days after emergence have only one tenth the level of activity that is present at emergence (HOROWITZ and FLING, 1955).

Dopa Decarboxylase (EC 4.1.1.26)

Dopa decarboxylase is a second enzyme of tyrosine metabolism, and it is likely involved in melanization and sclerotization of the cuticle. The enzyme was initially described by LUNAN and MITCHELL (1969). It has a pH optimum of 7.0, and is most active when dihydroxyphenylalanine (dopa) is used as substrate. Slight activity is observed using tyrosine. The enzyme shows no activity towards phenylalanine or lysine. MCCAMAN *et al.* (1972) have recently developed a micro assay which can be used to measure dopa decarboxylase and other aromatic amino acid decarboxylases in microgram quantities of tissue. The method is based on the extraction of the radioactive amine product of the decarboxylation reaction into an organic phase consisting of bisdiethylhexylphosphoric acid dissolved in chloroform. The organic phase can then be counted for quantitation of the product.

The levels of dopa decarboxylase during development have been reported by LUNAN and MITCHELL (1969) and by MCCAMAN *et al.* (1972) (Fig. 29). A small rise in the amount of activity per organism occurs at the second larval molt. This is followed by a decline, and then a large increase during late third instar. This rise reaches its maximum at pupation. Following pupation the amount of activity declines. Prior to emergence, activity levels again rise, and reach a maximum level which is approximately twice the level found at pupation. It is important to note that these rises occur around the times when the cuticle is being formed or tanned. This developmental profile is consistent with the proposed role of dopa decarboxylase in formation of a finished cuticle. This role is substantiated by the dissection experiments of LUNAN and MITCHELL (1969), who found that activity is distributed uniformly along the anterior-posterior axis. Furthermore, the bulk, if not all, of the activity is in the cuticle-epidermal parts of the animals.

Dopamine-N-Acetylase (EC 2.3.1.5)

The first measurements of dopamine acetylase have recently been reported by HODGETTS and KONOPKA (1973). Conversion of dopamine to N-acetyldopamine was catalyzed by extracts of larvae, pupae 90 hrs after the white puparium stage, in newly emerged adults and adults one day after emergence. In these *in vitro* assays, acetyl CoA served as the acetyl donor. Of the stages analyzed, the highest activity was found in one day old flies, when data are expressed on a live weight basis. Dopamine acetylase activity in extracts of the mutant *ebony* are similar to wild-type. HODGETTS and KONOPKA also determined the fate of radioactive tyrosine, dopa and dopamine when injected into animals of various ages. These data are used to formulate hypotheses concerning the bases of the body color mutants *black* and *ebony*.

Amylase (α-1,4-glucan-4-glucanohydrolase, EC 3.2.1.1)

The amylases of *Drosophila* have been the subject of rather thorough and careful genetic, biochemical, and developmental studies. The outcome of this work is a well-characterized gene-enzyme system that has already yielded some very interesting results, and promises to provide some important insights into the regulation of gene function in eukaryotes. The work on this system could well serve as a model for much work on other gene-enzyme systems. Much of the work on amylase has been summarized elsewhere (DOANE, 1969b).

The Biochemistry of Amylase

The reaction catalyzed by amylase is the hydrolysis of α-1,4-glucosidic linkages in molecules like starch and glycogen. Amylases are subdivided into α-amylases, which hydrolize internal glucoside bonds in molecules of three or more glucose units, and β-amylases which remove maltose residues from the non-reducing ends of polysaccharides. According to the characterization by DOANE (1969a), the amylases found in *Drosophila* are α-amylases, resembling in this respect the mammalian salivary and pancreatic enzymes.

Quantitative assays have been based on the measurement of the rate of hydrolysis of soluble starch, quantitated by the starch-iodine reaction after a fixed period of incubation (ABE, 1958; KIKKAWA, 1960; DOANE, 1969a), or by measurement of the free reducing ends produced by hydrolysis, using the colorimetric reaction with 3,5-dinitrosalycilic acid (DOANE, 1969b). The latter method is regarded as more accurate, since it avoids a number of known difficulties of the starch-iodine test and measures directly the number of reducing ends, and hence the number of bonds hydrolyzed.

Detection of amylase in electrophoretic gels continues to be based on the starch-iodine reaction. The original method (KIKKAWA, 1964) involved soaking the gel in a starch solution after the run, and then reacting with iodine solution. In a modification introduced by DOANE (1965, 1967a), the electrophoretic gel is incubated in contact with a second gel containing starch, and the latter gel is stained with iodine. In both cases, amylase activity appears as pale or clear spots in a dark background. DOANE (1967a) showed that, with appropriate choices of gel thickness and starch concentration, her contact method could be quantitated by scanning the spots with a densitometer. She also cut out and eluted spots to show that there is no appreciable loss of activity during electrophoresis, at least in acrylamide.

DOANE (1969a) has reported a considerable amount of biochemical characterization of *Drosophila* amylases. No purification had been attempted at that time, and the work was done on crude extracts. Amylase activity is stimulated by chloride ion, with an optimum between 0.025 M and 0.1 M. The several electrophoretic variants known (see genetics section) respond in a parallel fashion. The response to pH was also measured and the optimum found to be near 7.4, again with various electrophoretic forms being similar. Starch, amylopectin, β-limit dextrins, glycogen and amylose all serve as substrates. Starch and amylopectin give very similar specific activities, β-limit dextrins and glycogen yield slighlty lower activities in that order, and amylose is a much poorer substrate than the others. Strains carrying different electrophoretic variants of amylase often vary dramatically in total activity. If a series of strains is ranked with respect to activity with starch as substrate, the same rank order is found for the other substrates except amylose, the poorest substrate. The amylase activity in all strains is inhibited by EDTA and by the α-amylase inhibitor from wheat grain. The concentration of inhibitor and time of pre-incubation needed to produce a given level of inhibition varies between strains, roughly paralleling the specific activity in the respective untreated extracts. The most active strain was completely inhibited by EDTA at 0.05 M with a ten minute preincubation, or by a 0.01 M EDTA and a 20 min pre-incubation. Wheat grain α-amylase inhibitor completely inactivated the most active extracts with 0.1 mg/ml and a 10 min pre-incubation. Glutathione also acts as an inhibitor, giving complete inactivation at 0.05 M, but can also cause a transient activation of the enzyme at concentrations around 0.001–0.005 M with no pre-incubation. No inhibition by PCMB was demonstrable.

The molecular weight of amylase has been estimated on the basis of comparisons of the mobilities observed in acrylamide gels at different concentrations. The results indicated a value of about 50000, near the common molecular weight for α-amylases in other animals (DOANE, 1969b). The same data indicate that the different electrophoretic forms are of similar size.

Drosophila amylase is stable for at least 2 hrs at 25 °C, and for a year or more at −20 °C. Repeated freezing and thawing does not reduce activity, but some activity is lost at 4 °C overnight, and much is lost within 4–5 days at this temperature. This relative stability and the ease of assay make amylase a good candidate for purification.

Recently, DOANE *et al.* (1973, 1974) have reported a purification of amylases from *D. melanogaster* and *D. hydei.* They employ specific binding of the enzyme to glycogen to separate it from the bulk of other proteins. This single step gives over 200-fold purification with a yield of 67%. Final purification was achieved by disc electrophoresis, with the amylase band being cut out and eluted to recover the enzyme. This purified product appears to be homogenous. Analysis on SDS-acrylamide gels indicates that amylase contains a single polypeptide with a molecular weight of approximately 54500 (DOANE and KOLAR, 1973). The availability of purified amylase should make more detailed biochemical studies possible (see DOANE *et al.*, 1974).

Genetics of Amylase

Amylase was one of the first *Drosophila* enzymes without an externally visible phenotypic effect to be studied genetically. ABE (1958) and KIKKAWA (1960) found that various strains of *D. melanogaster* differ sharply in total amylase acitvity. The highest and lowest activity strains are sufficiently different to permit genetic analysis. F_1 progeny from a cross between high and low activity strains have an intermediate level of amylase activity. Both males and females show this intermediate level and the result is the same in reciprocal crosses, indicating an autosomal locus. The gene controlling these differences was mapped roughly to 2–80± (KIKKAWA, 1960).

With the introduction of methods that permitted the visualization of amylase following electrophoresis in agar gels (KIKKAWA, 1964), or in acrylamide gels (DOANE, 1965, 1967a), a series of electrophoretic variants was found in *D. melanogaster*. A total of seven positions may be occupied by major or minor bands in various strains. Bands are numbered from anode to cathode and *Amylase (Amy)* alleles are identified by superscript numbers. Homozygous strains may contain one or two major bands of activity and an equal number of minor or "shadow" bands, each located one position cathodal to a major band. Since the minor bands are not genetically independent, only major bands are named in designating alleles. Since two major bands can be present in homozygous strains and these patterns segregate as a unit in ordinary genetic experiments, it is concluded that the two major bands are coded for by distinct, but closely linked loci. The *Amy* alleles in these strains are designated by using two superscripts separated by a comma. Major bands are found only in positions 1, 2, 3, 4, and 6. Positions 5 and 7 are only occupied by minor bands, so there are no alleles carrying these superscripts.

KIKKAWA (1964) found seven different homozygous patterns (not to be confused with the seven banding positions). DOANE (1969a) confirmed these, and added one more. The naturally occurring patterns identified thus far (DOANE, 1969a, b) include Amy^{1}, $Amy^{1,2}$, $Amy^{1,3}$, $Amy^{1,4}$, $Amy^{2,6}$, $Amy^{1,6}$, $Amy^{3,6}$, and $Amy^{4,6}$. In addition, BAHN (1968) recovered certain recombinants during a fine structure analysis of the *Amy* region (see below), adding Amy^{2}, $Amy^{2,3}$, $Amy^{4,3}$, and an independently generated $Amy^{2,6}$ to the list of known patterns. Heterozygotes formed by crossing strains with any two of these patterns have a pattern of major and minor bands that is a simple sum of the parental patterns with no added hybrid bands.

The electrophoretic variants have been used to refine the genetic mapping data for the *Amy* loci. KIKKAWA (1964) obtained a position of 2-78.1, DOANE (1963, 1969a) found a value of 2-77.3 and BAHN (1968) found 2-77.9 and 2-78.4 with two independent sets of data. These values agree within the probable experimental error. DOANE (1967b, 1969a) and BAHN (1971) have also attempted cytogenetic localization of the *Amy* locus. The former author places it to the right of, but near, salivary chromosome band 52F. The latter author, using an extensive series of translocations, narrowed this down to a position very close to band 54A.

Electrophoretic variants have also been used to map the *Amy* locus in *D. hydei* (DOANE, 1971). The locus is on the fifth chromosome in this species between *cn* and *vg*. This position is homologous to the location in *D. melanogaster*. Using the *cn* and *vg* markers to identify putative X-ray induced deficiencies and other chromosomal abnormalities, the cytogenetic map position has also been determined (DOANE, 1971, DOANE *et al.*, 1974).

BAHN (1968) has obtained direct evidence that *D. melanogaster* strains that have two major amylase isozymes also contain two separable (but closely linked) genes coding for the electrophoretically distinguishable forms. He used two closely linked outside markers, *curved* (*c*, 2-75.5) and *welt* (*wt*, 2-82), and examined chromosomes recombinant between these two markers. Thus, for example, heterozygous *c* Amy^1 *wt*/ + $Amy^{2,3}$ + females are mated to homozygous *c* Amy^1 *wt* males. Single male progeny that are homozygous for only one of the outside markers are mated to females with appropriate balancers to preserve the recombinant chromosomes. The individual recombinant males are then recovered and analyzed electrophoretically. In this example, recovery of males with only bands one and two or only one and three indicates separation of Amy^2 from Amy^3, and hence recombination between these *Amy* loci. Among 5039 chromosomes recombinant between *c* and *wt* (representing an estimated total of over 77000 offspring), six were recombinant between two *Amy* genes. The estimated map distance separating the *Amy* sites is .008 units. On the basis of the associated outside markers, the order of the *Amy* loci can be inferred. BAHN concludes that Amy^3 and Amy^6 lie to the right of Amy^1, Amy^2, and Amy^4. DOANE (1967b) has found rare apparent recombinants in a cage population, supporting the conclusion that there are at least two separable *Amy* loci.

The various electrophoretic patterns found in homozygous strains are associated with dramatically different total amylase activities. The differences account for the high and low activity strains initially identified by ABE (1958) and KIKKAWA (1960). DOANE (1969a,b) has also found considerable variation between strains with only band 1. She has added a letter to the superscript of these alleles to indicate this difference. The order of enzyme activity from highest to lowest is as follows: $Amy^{2,6}$, $Amy^{1,6}$, $Amy^{3,6}$, $Amy^{4,6}$, $Amy^{1,2}$, $Amy^{1,3}$, $Amy^{1,4}$, Amy^{1-a}, Amy^{1-b}, and Amy^{1-c}. As mentioned previously, the rank order is the same for all of the four good substrates. It can be seen that all of the strains with high activity have the band 6 isozyme in combination with some other form, and the lowest activity strains have only band 1. The range of activity from $Amy^{2,6}$ to Amy^{1-c} is about seven-fold, and the range from Amy^{1-a} to Amy^{1-c} is about two-fold. DOANE (1969b) has also used her quantitative gel scanning method to compare the relative contributions of the different isozymes in flies with two major forms. In general, the more cathodal band contributes a higher proportion of the total activity. There are sex specific differences in the exact figures, which may be related to the fact that there are also tissue specific differences (see development section).

Heterozygotes produced by crosses between any two of the above mentioned strains have an intermediate level of amylase activity, although not always precisely halfway between the levels in the parental strains (DOANE, 1969a, 1969b). The relative contributions of two bands inherited from one parent are the same in the heterozygote as in the parent, but with the absolute values about half as great (DOANE, 1969b). Thus, the amylase activity in the heterozygote is approximated both quantitatively and qualitatively by the average of the two parental pattern.

There is considerable evidence that suggests the existence of genetic variability at the *Amy* loci beyond that made apparent by the varied electrophoretic patterns.

We have already mentioned the differences in total amylase activity among strains with a single amylase band at position 1. DOANE (1969b) has also found that the *Amy*^{2,6} line recovered by BAHN (1968) during his recombination studies differs from the original *Amy*^{2,6} of KIKKAWA (1964) in both total activity and the relative distribution between the two isozymes. She has also found that isozymes occupying the same position in electrophoretic gels, but derived from different strains, may differ in heat stability.

The most frequently observed pattern in *D. melanogaster* is *Amy*^{1} alone (KIKKAWA, 1964; DOANE, 1969a), and this has been regarded as the primitive form. Surveys of other *Drosophila* species (summarized by DOANE, 1969b) reveal that polymorphisms are common, but in all cases appropriately tested, homozygous individuals contain only one isozyme. Thus, *D. melanogaster* is the only established case where closely linked, duplicate *Amy* loci have been found. The duplication is therefore presumed to be of relatively recent origin. It would be interesting to know whether the *Amy*^{1} strains carry only the original, unduplicated locus or duplicate but still identical genes. Both situations might obtain, and this might explain the differences in total activity between *Amy*^{1} strains.

Amylase has been used as a marker in one of the few studies that has provided direct evidence for the idea of co-adapted gene blocks associated with chromosomal inversions (PRAKASH and LEWONTIN, 1968). The phylogeny of a series of inversions in the third chromosome of *D. pseudoodscura* is fairly well-established. *Amy* is one of two known polymorphic loci in this region. The allele frequencies associated with each inversion pattern were distinct, but fairly constant within each inversion pattern over the entire geographic range of the species. This is taken as evidence for real and significant genetic differences between chromosomes with different inversion patterns, probably maintained by selection.

Amy is possibly the only enzymatically defined locus at which a classic position effect variegation has been observed. BAHN (1971) obtained a translocation that places *Amy* in the heterochromatic region of the X chromosome. The variegated position effect was detected in the form of varied intensity of the *Amy* band stained in agar gels. Many of the classic features of position effect variegation at other loci (see BAKER, 1968) were observed.

Development and Regulation

Observations on changes in total amylase activity with development have been reported by ABE (1958), KIKKAWA and ABE (1960) and DOANE (1965, 1969b). The activity is low in newly laid eggs, and begins to increase shortly before hatching. There is a steady increase during the larval stages, except for a temporary dip at each molt. The most rapid increase occurs during the third instar, leading up to a maximum just before pupation. Enzyme activity drops sharply at pupation to around 15% of the peak value, and then continues to decline slowly until adult emergence. The first 3–4 days after eclosion are marked by another fairly sharp increase up to a steady level only slightly lower than the third instar peak. The various strains differ in detail, but are qualitatively similar (DOANE, 1969b).

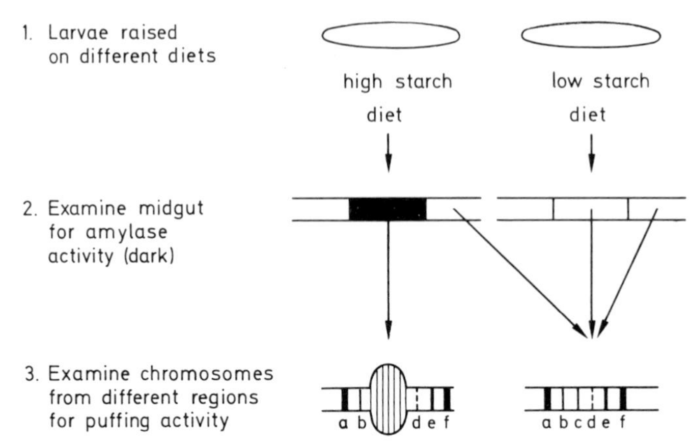

Fig. 30. Correlation of chromosomal puffing activity with presence of a gene product. At step 1, larvae are raised on high or low starch diets. In step 2, only those larvae raised on starch have a concentrated band of amylase activity in the midgut. In step 3, chromosomes from cells in different regions of the gut and from larvae raised on different diets are examined. The lines marked a–f represent chromomeres visible on the polytene chromosomes. One chromomere (band c) is shown puffed only in those cells containing amylase activity. Independent genetic evidence is used to determine the region of the chromosome containing the gene for amylase (see text)

The same authors have examined the tissue distribution of the enzyme. On the basis of dissection studies, most amylase activity is found in the midgut and hemolymph in both larvae and adults, although traces are found in the salivary glands, fat body, Malpighian tubules and ovaries (DOANE, 1969b). It is not known whether the amylase found in these tissues arises *in situ* or is transported there by hemolymph, which does contain a considerable amount of the enzyme. It is most interesting that the relative contributions of different amylase isozymes, in those strains having more than one, vary with developmental stage and from tissue to tissue. This suggests that the two closely linked *Amy* loci may not be under coordinate control, although it is impossible to say at what level the differences might be introduced.

DOANE (1969b, 1971) has further refined the localization of amylase in the gut by histochemical methods. In *D. melanogaster,* the activity is confined primarily to the posterior midgut and to a smaller, less intense region in the anterior midgut.

ABE (1958) reported that addition of starch to the diet on which flies were raised increased the activity of amylase in the adults. DOANE (1969b) has confirmed this effect. The response is most noticeable in males, and is suppressed when sucrose is also added to the medium. Again, the relative contributions of different isozymes may vary with the food.

DOANE (1969b, 1971) and DOANE *et al.* (1974) have been able to combine knowledge about the cytogenetic locus of the *Amy* gene, the tissue specificity of its expression, and the response to dietary conditions to produce a remarkable correlation between apparent genetic activity, (as reflected in chromosomal puffs), and the specific presence of the corresponding gene product. She has turned to *D. hydei* for this work because the cells in the midgut of this species have polytene chromosomes of sufficient size to permit good cytogenetic work. She used X-ray induced rearrangements to determine the probable cytogenetic locus of *Amy* on the fifth chromosome. She also showed a tissue specific localization to certain regions of the midgut, and a response to starch similar to that seen in *D. melanogaster*. Examination of the appropriate chromosomal region in cells in different parts of the midgut, and in flies raised on a starch or sucrose medium, reveals a puff that is present only in cells that are known to have a high level of amylase activity (Fig. 30). While some additional work may be necessary to establish definitively the relationship between the puff and the enzyme, this work is a beautiful illustration of what may be hoped for when a considerable body of genetic, biochemical and developmental information is brought to bear on one system.

Non-Specific Hydrolytic Enzymes

Amino Peptidases and Proteases

We shall consider together here a group of enzymes that, in *Drosophila,* is poorly defined with respect to substrate specificity. Most of the work has taken advantage of artificial substrates, such as L-leucyl-β-naphthylamide, using diazonium salts to trap the naphthol released by enzymatic cleavage (BECKMAN and JOHNSON, 1964c). A few studies have used natural proteins as substrate, and determine general proteolytic activity in terms of amino acids (e.g., tyrosine) rendered acid soluble (KIKKAWA, 1968).

Biochemistry

Relatively little biochemistry has been done on these enzymes. It appears that there are several different peptidases and proteases, and that at least some of them have broad specificity. This, of course, renders further biochemical definition of the system difficult, and also complicates the interpretation of developmental studies on total proteolytic activity.

BECKMAN and JOHNSON (1964c) report at least six distinct zones of activity in extracts of *D. melanogaster* electrophoresed in starch gels and stained with L-leucyl-β-naphthylamide. They term these bands of activity leucine amino peptidase (Lap) A–F. SAKAI *et al.* (1969) tested a variety of other amino acid-β-naphthylamides as substrates for BECKMAN'S and JOHNSON'S *Lap D,* and showed that this enzyme had wide specificity. They propose designating it simply as an amino peptidase. PATTERSON and LANG (1954) measured activity of salivary gland extracts in hydrolysis of leucineamide, leucylglycine and leucylglycylglycine in various reaction mixtures. They found that the ratios of activities toward the three substrates varied with ion composition (Mn^{++}, Ca^{++}, and Fe^{+++}), presence or absence of citrate or cysteine, and in response to pre-extraction with acetone. They interpret this to mean that multiple peptidases are present. KIKKAWA (1968) reached a similar conclusion for general protease activity. His experiments with trypsin inhibitor suggest multiple proteases, some resistant to the inhibitor.

In contrast to *D. melanogaster, D. busckii* has a single electrophoretic zone of Lap activity (JOHNSON *et al.,* 1967). It would seem desirable to examine the patterns obtained with other substrates to see whether this indicates fewer enzymes or greater specificity.

Genetics

BECKMAN and JOHNSON (1964) report electrophoretic variants for Lap zones A and D. The *Lap D* variants behave as if controlled by a single pair of codominant autosomal alleles. Heterozygotes contain both parental forms and no hybrid band. SAKAI *et al.* (1969) report a third variant *(Lap D^x)* also inherited as a simple codominant. They map the gene roughly to 3-100$\pm$2.5, in general agreement with BECKMAN'S and JOHNSON'S earlier crude positioning between 90 and 110 on chromosome III.

Lap A has, in addition to two electrophoretic variants, an apparent null allele (BECKMAN and JOHNSON, 1964c). Presence of the enzyme is, not surprisingly, dominant to absence, but it is not known whether the heterzygote has a reduced amount of enzyme. The null allele at *Lap A* can be present in combination with either the slow or fast variant at *Lap D,* and the two loci appear to be closely linked. The authors suggest origin of the two loci by gene duplication. The *Lap A^f/Lap A^s* heterozygotes contain a rather diffuse band intermediate in position to the parental forms, and it is difficult to determine whether the parental forms are also present.

The single band of Lap present in *D. busckii* (JOHNSON *et al.*, 1967) has three known electrophoretic alleles inherited as codominants with no hybrid enzyme forms. However, crosses of F_1 heterozygotes give non-Mendelian segregation ratios deficient in homozygotes. It appears that heterozygotes enjoy a selective advantage in this system (RICHARDSON and JOHNSON, 1967).

KIKKAWA (1968) examined 60 strains of *D. melanogaster* for quantitative differences in proteolytic activity. High and low activity strains were selected that differed about two-fold in specific activity. The genetic basis of the difference was not fully investigated.

The classical developmental mutant *lethal-meander* (*lme*, 2–71 to 73) may have a deficiency in proteolytic enzymes in the digestive system (HADORN, 1956; CHEN and HADORN, 1955). Flies homozygous for this mutation arrest early in the 3rd instar, and never enter metamorphosis. Starvation of normal larvae produces a good phenocopy. Chromatographic studies of *lme* larvae show them to be extremely deficient in free amino acids and small peptides. Total intake of food is normal in the mutants, but feeding them casein fails to get free amino acids into the hemolymph. However, free amino acids in the food do get into the hemolymph, so the block does not appear to be in transport. Finally, an extract of the gut of *lme* incubated with casein produced much less free amino acid (analyzed chromatographically after incubation) than did an extract of wild-type gut. Reduction of proteolytic activity in the larval gut of *lme* was confirmed by MEYER-TAPLICK and CHEN (1960) and by WALDNER-STIFELMEIER (1967). One should not, however, conclude that reduction of proteolytic activity is the only, or even primary, effect of *lme*. For a discussion of other effects of this mutant, see CHEN (1971), pp. 153–156.

Development

The zones A–F of Lap in *D. melanogaster* appear in a stage specific pattern (BECKMAN and JOHNSON, 1964). Zone A is present at all stages, while B–F are primarily pupal enzymes. Zone D appears earlier than the others, and indeed, SAKAI *et al.* (1969) report that *Lap D* is detectable at low levels in larvae and adults. The appearance of new peptidase bands in pupae may be associated with the histolysis characteristic of this period. However, quantitative estimates of total proteolytic activity show the highest activity in larvae, and a low point in early pupae (KIKKAWA, 1968; WALDNER-STIFELMEIER, 1967). KIKKAWA'S measurements, based on release of tyrosine from casein, show the activity/individual to be 10-fold higher in larvae than adults, and early pupae are another 4-fold lower than adults.

Females have a greater activity than males, even when adjustment is made for body weight. KIKKAWA suggests that the best correlation of total activity is with gut size, both in explaining stage differences and sex differences. The high and low activity strains isolated by KIKKAWA had similar developmental patterns and similar sex differences.

Comparison of the activity in isolated gut to whole extracts suggests the presence of protease inhibitors in *Drosophila* body fluids. The stage and sex specific differences in activity measured in whole extracts cannot, however, be attributed to variations in the level of inhibitor, nor can the high and low strains be explained in this way (KIKKAWA, 1968).

Esterases

The esterases represent a loosely related family of enzymes that are usually defined operationally by their hydrolysis of artificial substrates like α-naphthyl acetate. A staining reaction in electrophoresis gels is achieved by trapping enzymatically released naphthol (or a naphthol derivative) with a diazonium salt to produce an insoluble stain. It is difficult to determine the true physiological substrates for the various esterases, and many probably have broad specificity. Studies on *Drosophila* esterases have relied heavily on electrophoretic methods, and most of the enzymes remain poorly characterized biochemically. Esterases were among the first enzymes studied electrophoretically in *Drosophila,* and variants of Est-6 (WRIGHT, 1961, 1963) were the first reported case of electrophoretic polymorphism in this organism.

Biochemistry

Number of Esterases

Extracts of *D. melanogaster* electrophoresed (generally on starch gels) and stained for esterase reveal a minimum of six zones of activity (BECKMAN and JOHNSON, 1964b), and perhaps as many as ten (WRIGHT, 1963). WRIGHT numbered the zones, while BECKMAN and JOHNSON use letter designations A–F. Their zone D corresponds to WRIGHT'S Est-6, and the latter nomenclature is used because of WRIGHT'S priority. Correspondences among the other esterases are less easily established, and no consistent nomenclature has evolved. Other species of *Drosophila* show a similar range of esterase zones. McReynolds (1967) reports up to 12 zones of activity in mass homogenates of *D. virilis* and two sibling species, while Johnson *et al.* (1968) found six in *D. aldrichi* and *D. mulleri.*

Specificity and Inhibition

Further biochemical investigation of esterases has leaned heavily on differences in substrate specificity and sensitivity to inhibitors demonstrable in electrophoretic gels. Wright (1963) reports that bands 5, 6, and 9 in his nomenclature are not sensitive to eserine sulfate and, hence, apparently are not cholinesterases. He emphasizes that the physiological substrates are unknown, and that classification is essentially arbitrary. Johnson and co-workers (Johnson *et al.*, 1966; Johnson and Bealle, 1968) used specificity toward α and β naphthol esters and sensitivity to alcohols to classify esterases. Advantage is taken of the fact that α-naphthol—fast blue RR differs in color from the corresponding β-naphthol compound. The gel is sliced, and each half is stained with a mixture of α- and β-esters, with alcohol added to one half and not the other. The color of the band indicates whether α- or β-specificity predominates, and comparison of the color and intensity of the bands in the two slices indicates inhibitor sensitivity.

Johnson *et al.* (1966) give the following classification of major esterases:

Est-A — specific for α-ester, activity enhanced by propanol,
Est-C — β-specific, not affected by propanol,
Est-6 — βspecific in mixture. α-activity alone inhibited by propanol.

Classifications of this type are useful for comparative studies (see below), but leave us with little idea of the physiological role of the enzymes.

Species Comparisons

A number of papers have appeared dealing with homologies between esterases of different *Drosophila* species. Wright and MacIntyre (1963) report that *D. simulans* has an esterase with a mobility identical to Est-6 of *D. melanogaster,* and that identical electrophoretic variants are also found in both sibling species. The genetic mapping done on the variants (see Genetics section) indicates that the controlling genes occupy homologous positions in the respective genomes. The authors raise the interesting question whether the polymorphism existed in an ancestral species and was passed down to both, or arose independently after speciation. There is no obvious way to resolve this question. Triantaphyllidis (1973) believes that both Est-6 and Est-C are homologous in these two species.

Johnson *et al.* (1966) used the specificity and inhibitor tests mentioned above in an attempt to test esterase homologies in an extensive series of *Drosophila* species. The results are too extensive to consider in detail, but in broad outline, species belonging to a given recognized subgenus do tend to have homologous esterase patterns.

McReynolds (1967) indentified 3 zones of esterase activity in *D. virilis, D. americana* and *D. novamericana* that seemed to be homologous by position and specificity. One of these, designated Ev-4, formed a hybrid enzyme band in interspecific crosses, providing additional support for homology. Similarly, JOHNSON *et al.* (1968) proposed homology between most bands found in *D. aldrichi* and *D. mulleri*. Bands in corresponding positions showed similar response to various substrates and inhibitors. However, bands F and G of *D. aldrichi* are absent in *D. mulleri*. Band F is present only in *D. aldrichi* males, but is not sex-linked. In a second paper (KAMBYSELLIS *et al.*, 1968) the comparison was extended to the tissue distribution of the various esterases in these two species. Enzymes believed to be homologous by previous criteria do indeed have the same tissue specificity (see under development). *D. aldrichi* contains an ejaculatory bulb esterase missing in *D. mulleri*. JOHNSON and BEALLE (1968) have examined ejaculatory bulb esterases in 93 species of *Drosophila*. Species from the *Drosophila* subgenus have alcohol inhibited β-esterase in this tissue, while members of the *Sophophora* subgenus lack β-esterases here. When β-esterases not inhibited by alcohol are found, they are also present in the female. Alcohol inhibited β-esterases are never found in the female.

HUBBY and NARISE (1967) used the retention of the ability to form hybrid enzyme molecules as a measure of the conservation of molecular similarity. Since hybrid enzyme formation presumably requires rather extensive similarity in tertiary structure, the authors argue that this is a more significant indicator than electrophoretic similarity. They used substrate specificity etc., to identify, in various *Drosophila* species, esterases thought to be homologous to Est-5 of *D. pseudoobscura*. The esterases from closely related species could form *in vitro* hybrids. Electrophoretic variants from a given species often differed in their ability to hybridize with esterase from another species. *In vitro* hybrids would not form between esterases of distantly related species. In contrast, malate dehydrogenase from all tested species of *Drosophila* formed *in vitro* hybrids. This is interpreted to mean that malate dehydrogenase has been more conservative during the evolution of *Drosophila*.

Purification

One esterase from *Drosophila* has been purified. NARISE and HUBBY (1966) purified esterase-5 from *D. pseudoobscura*. The purification employed ammonium sulfate fractionation, Sephadex G-100 chromatography, DEAE Sephadex A-50 chromatography, acetone precipitation and preparative acrylamide gel electrophoresis. Purification of the final pooled fraction was 66-fold, and recovery was 2.5%. The peak fraction was 310-fold purified. Actual purification and yield are presumably higher, since the crude extract contained several other esterases which were removed during purification. The purified product showed greatest activity toward β-naphthyl acetate, and had a pH optimum at 6.5. Molecular weight, estimated by velocity gradient centrifugation, was 107000. No significant differences in biochemical properties were noted for allelic variants with very different electrophoretic mobilities (0.85 and 1.12 relative mobilities).

Genetics

While the number of esterase zones reported in *D. melanogaster* ranges from 6 to 10, only two are well characterized genetically. Esterase-6 is characterized by electrophoretic variants that are co-dominantly expressed in heterozygotes with no evidence of an intermediate hybrid band. Standard mapping using recessive markers places the *Est-6* locus at 3-36.8 (WRIGHT, 1963). An apparently homologous esterase in *D. simulans* maps to 3-25.2, the homologous position in *D. simulans* (WRIGHT and MACINTYRE, 1963). WRIGHT and MACINTYRE (1965) also found strains of *D. melanogaster* with electrophoretically indistinguishable forms of Est-6^F that differed in heat stability. The differential heat stability segregated as expected for variants allelic at the site that controls the electrophoretic variants. This illustrates the fact that electrophoretic identity does not establish molecular identity. BELL *et al.* (1972) induced six null mutations at the *Est-6* locus with EMS. See p. 129 for an outline of their method. BECKMAN and JOHNSON (1964) report electrophoretic variants for two bands they designate C and D. Their *Est-D* apparently corresponds to WRIGHT'S *Est-6*. *Est-C* also is expressed codominantly with no hybrid band. It is linked to *Est-6* (hence on 3rd chromosome), and is closely linked to the gene controlling larval alkaline phosphatase. JOHNSON (1964b) has also reported an inherited deficiency for *Est-C*, which is inherited as a simple recessive. Allelism to the electrophoretic variants does not appear to have been tested. MIZIANTY and CASE (1971) have found an apparent null allele for *Est-A*. Presence of the enzyme is dominant to absence, but heterozygotes seem to have reduced activity. Segregation is consistent with control by a single pair of alleles.

Genetic variants have also been reported in esterases found in other *Drosophila* species. NARISE and HUBBY (1966) report that *esterase-5* of *D. pseudoobscura* has at least 6 alleles, and is sex-linked. MCREYNOLDS (1967) examined esterases in three related species from the *virilis* group of *Drosophila*. Up to 12 zones of esterase activity were seen, but only three, designated Ev-1, Ev-4, and Ev-5 were sufficiently dark to study in single flies. Ev-4 had two alleles that formed a hybrid band in heterozygotes, and Ev-5 had 5 alleles. In test crosses with recessive markers on all chromosomes, backcross progeny heterozygous for chromosome 2 were the only ones also heterozygous at both esterase loci. Hence, the electrophoretic loci for both enzymes are linked on the 2nd chromosome. There was 10% recombination between the two loci.

Physiological Significance of Electrophoretic Variants

The selective significance of electrophoretic variants and the nature of the mechanisms that maintain polymorphisms have been of concern primarily to population geneticists. However, the use of electrophoretic markers in developmental or physiological studies is sometimes based on the assumption that flies carrying these markers are physiologically normal. Likewise, accurate genetic mapping is complicated if survival is not equivalent among various progeny classes. Hence, the question of selective neutrality is of concern to workers in other fields. Esterase variants have been used widely in experimental approaches to this question.

MACINTYRE and WRIGHT (1966) found that cage populations of *D. melanogaster* established with allele frequenies of *Est-6* quite divergent from what is normally found, rapidly reached a stable equilibrium frequency close to that normally observed. However, when several generations were allowed for randomization of linkage groups prior to setting the starting frequencies, the results were less consistent. The same was true when the genetic background was controlled. Hence, it appeared that the major selective force in establishing equilibrium operated at other linked loci. YARBROUGH and KOJIMA (1967) allowed randomization in a base population for 30 generations before pulling out pair matings to establish sublines. They determined the *Est-6* alleles in the sublines, and constructed test populations with the desired allele frequencies at *Est-6* by combining 20 such sublines. Again this procedure was intended to randomize linkage to other genes. Populations started with *Est-6*^{F} frequencies of 0.9 and 0.1 were established. They converged to a range between 0.25 and 0.45 (the base population was stable at about 0.3). Convergence was relatively rapid for 15 generations. This result suggests that there is real selection for a balanced polymorphism at this locus.

In a second paper KOJIMA and YARBROUGH (1967) tested the mechanism of selection. Classical heterozygote superiority can maintain balanced polymorphism, but the genetic load involved in maintaining polymorphism independently at many loci seems intolerable (LEWONTIN and HUBBY, 1966). KOJIMA and YARBROUGH found that fecundity (egg lay) of females of the three genotypes could not explain the equilibrium frequencies. They therefore examined egg—adult viability. Eggs of known genotype were collected and mixed in proportions simulating the expectation from populations with *Est-6*^{F} frequencies of 0.7, 0.5, 0.3, and 0.15. They then examined the yield of adults of each genotype after the eggs were reaised in competition. The population set up simulating normal equilibrium (0.3) showed no significant differences in viability. However, whenever any genotype was present in excess, it had greatly reduced viability. Below its equilibrium level, each genotype enjoyed greater viability. Based on this work, the authors propose a model of balanced polymorphism based on frequency dependent selection. A given genotype is assumed to be selected against when it exceeds its equilibrium frequency, but at equilibrium there is no selection against any genotype. One possible mechanism that can explain this pattern of selection is differential exploitation of different micro-environments. If one assumes that the different alleles allow utilization of slightly different resources within the environment, all genotypes can do equally well when they are in balance with the availability of the relevant resources.

YAMAZAKI (1970) looked for evidence of fitness differences at the *Est-5* locus in *D. pseudoobscura*. Two major electrophoretic variants of this sex-linked locus were extracted from a long standing cage population, in which randomization of linkage should have occurred. The cage maintained a stable polymorphism. He set up cages at various allele frequencies and also examined viability, development time, fecundity, and possible frequency dependent selection. He was unable to detect a measurable difference in fitness in any of these tests. He concludes that the results are consistent with selective neutrality or very weak selection. PRAKASH and MERRITT (1972) used alleles of *Est-5* and *Acph-6* in *D. pseudoobscura* to obtain direct evidence for co-adapted gene blocks associated with inversions. The so called sex-ratio (SR) and standard (ST) chromosome differ by three inversions in the right arm of the X chromosome. The various alleles of the *Est-5* and *Acph-6* loci show a strongly non-random association relative to the SR and ST arrangements. Comparing chromosomes with one arrangement or the other, there is no evidence of significant allele frequency differences over a wide geographical range. TRIANTAPHYLLIDIS (1973) found that the frequencies of apparently identical alleles in *D. me-*

lanogaster and *D. simulans* were markedly different in sympatric populations. It is suggested that the two species are subject to different selection forces. However, the assumption that enzymes in the two species with identical electrophoretic mobilities are really identical is questionable, particularly in view of the differences in heat stability of electrophoretically indistinguishable esterase-6 variants in *D. melanogaster* reported by WRIGHT and MACINTYRE (1965).

Development and Tissue Specificity

Very little work has been done on the developmental histories of the various esterases in *Drosophila*. KAMBYSELLIS *et al.* (1968) examined developmental changes and tissue distributions in *D. Aldrichi* and *D. mulleri* in connection with establishing homologies. They note that there are changes with aging in adults, and that there is a male specific band that appears about five days after eclosion (Fig. 31). By dissection and electrophoresis they found no detectable esterase in ovaries, testes, scutellum, halteres, legs, fat body, pericardial cells, epidermis and cuticle, wings, thoracic muscle, and salivary glands. Esterase bands A–F in *D. aldrichi* were found distributed in various tissues as follows (see also Fig. 31). Bands E and E′ are primarily in nervous tissue, B is specific to proboscis and antennae, A and A′ are specific to antennae. The hemolymph is rich in D, and the gut has only C. Est-F is the male specific band, and is found in the ejaculatory bulb. Freshly mated females may contain a trace of this band. A band not detected in whole extracts is found in paragonia, and is designated Est-P. A second strain of *D. aldrichi* had essentially identical tissue distributions, but *D. mulleri* lacked both the ejaculatory bulb esterase and the Est-P.

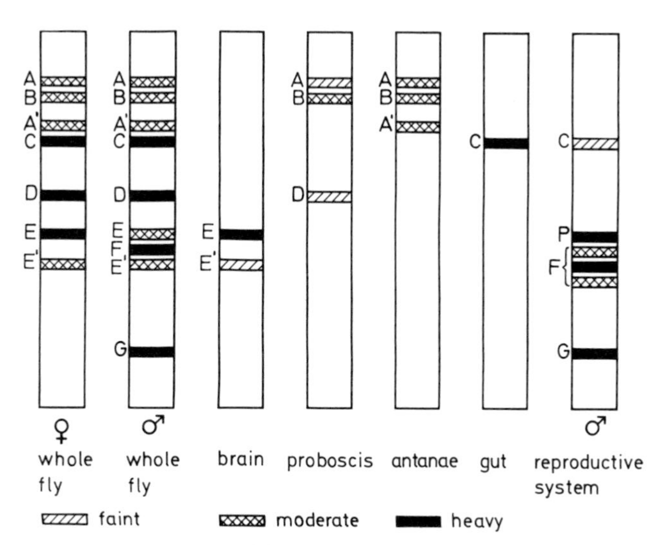

Fig. 31. Diagrammatic representation of the sex and tissue specificity of esterases in *Drosophila aldrichi*. The tissues are indicated under each column, and the letters identify the bands. Esterases B, C, E, E′, and P are α specific, and A, A′, D, F, and G are β specific. (Redrawn from data of KAMBYSELLIS *et al.*, 1968)

D. melanogaster also shows some sex differences in esterase pattern (JOHNSON, 1964a). Est-6 has consistently higher activity in males, and two other bands have altered mobility. In addition, males have one extra band. XO males show all of these characteristics, so the differences are sex-limited, not Y-linked.

A possible role of some esterases in juvenile hormone metabolism has been suggested in other insects (WHITMORE *et al.*, 1972). This makes developmental and genetic studies on esterases even more interesting, but no indication of such a relationship has been demonstrated in *Drosophila* as yet.

Alkaline Phosphatases

Alkaline phosphatase (Aph) constitutes an operationally defined group of enzymes that are capable of hydrolyzing α-naphthyl-phosphate (and related artificial substrates) at alkaline pH. The true substrates and physiological roles of various enzymes in this group are essentially unknown, and it is likely that the various "isozymes" are quite unrelated in function.

Biochemistry

The number of electrophoretically distinct alkaline phosphatase bands reported in various *Drosophila* species ranges up to seven in *D. melanogaster* (SCHNEIDERMAN *et al.*, 1966), seven in *D. pseudoobscura* (HUBBY and LEWONTIN, 1966), and four in *D. ananassae* (JOHNSON *et al.*, 1966). As we shall see below, not all of these bands necessarily represent distinct gene products. Biochemical characterization is quite limited. SCHNEIDERMAN *et al.* (1966) attempted to look at substrate specificity by comparing the relative staining intensities of the seven bands with various substrates. Bands 2–7 have activity with ATP, AMP, and glucose-6-phosphate with no apparent specificity. Band 1 was too faint to study.

HARPER and ARMSTRONG (1972) have done more extensive work on a very active larval Aph previously characterized genetically by BECKMAN and JOHNSON (1964a). They partially purified and characterized two allelic forms designated Aph-4 and Aph-6. The quantitative assay employed p-nitrophenyl phosphate as a substrate at pH 8.0. They also tested other substrates by following release of inorganic phosphate. The purification involved extraction with n-butanol, precipitation with $(NH_4)_2SO_4$, and chromatography on sepharose 6-B and DEAE-cellulose. The final product, representing a 35% yield, was 277-fold purified.

Both forms of the enzyme had a narrow pH optimum. The optimum was at pH 8.0 for Aph-6 and at 8.5 for Aph-4. Both were inhibited by CN^- and inorganic phosphate at 10^{-3} M, and were slightly inhibited by cysteine. They were not sensitive to EDTA or deoxycholate in the assay medium. However, dialysis against EDTA produced an inactive apo-enzyme whose activity is largely restored by Zn^{++}.

The two electrophoretic forms were quite similar in substrate range, and in all tested properties other than the minor difference in pH optima. Unlike many alkaline phosphatases (e.g., that in *E. coli*), the enzyme is not particularly heat stable. The enzyme had a wide substrate specificity, the best substrates being o-phosphotyrosine, o-phosphothreonine, o-phosphoserine, o-phosphoethanolamine, β-glycerophosphate, 5′-AMP, phenolphthalein monophosphate, histidinol phosphate, phosphocholine and α-glycerophosphate. Others that were only moderately good substrates include ADP, ATP, and some sugar phosphates. This list dramatizes the difficulty of determining the physiological substrates of alkaline phosphatases.

The authors suggest that the normal function of this larval Aph may be conversion of o-phosphotyrosine present in large amounts in late third instar larvae, to tyrosine which is used in puparium formation. However, JOHNSON (1966) has reported a mutant (*Aph*^{o}) deficient in this enzyme, and he gives no indication that puparium formation is in any way abnormal. Perhaps more careful attention to this possibility is in order.

Genetics

The earliest report of genetic variants of an alkaline phosphatase in *Drosophila* is that of BECKMAN and JOHNSON (1964a). They noted two forms of the major band detectable in late third instar larvae (the subject of the biochemical work of HARPER and ARMSTRONG). The variants are designated *Aph*^{F} (fast) and *Aph*^{S} (slow) in this paper. The two parental forms and a new hybrid band are present in heterozygotes. The genetics indicate simple Mendelian codominant segregation. Taking advantage of the fact that variants at the *Est-6* locus can be detected in one half of a sliced starch gel and *Aph* in the other half, the *Aph* locus was mapped relative to *Est-6* and found to be closely linked. Hence, larval Aph is controlled by a gene on chromosome III in *D. melanogaster*. JOHNSON (1966b) has also reported an apparent null allele of this same gene. An inbred strain of *ca*^{r} completely lacked the larval Aph band. The variant was designated *Aph*^{o}. The *Aph*^{F}/*Aph*^{o} heterozygote had only the fast band. However, the *Aph*^{S}/*Aph*^{o} hetereozygotes contain an intermediate band in addition to the slow one. The intermediate band corresponded in position to the hybrid band found in *Aph*^{S}/*Aph*^{F} heterozygotes. The most attractive explanation of this observation is that *Aph*^{o} produces a polypeptide that is electrophoretically similar to *Aph*^{F}, but enzymatically inactive. However, one cannot exclude the possibility the *Aph*^{o} polypeptide cannot associate with *Aph*^{F}, or that the complex is inactive. Like the electrophoretic variants, *Aph*^{o} is linked to *Est-6*. This, together with the segregation pattern, supports the conclusion that *Aph*^{o} is allelic to *Aph*^{F} and *Aph*^{S}. WALLIS and FOX (1968) report similar observations with respect to a null allele of the same enzyme, and give a map position of 3–47.3, which agrees with BECKMAN'S and JOHNSON'S (1964) position relative to *Est-6*. MACINTYRE (1966a) finds a position of $3–46.3 \pm 0.5$. His mapping procedure nicely illustrates the methods that must be used when a larval enzyme marker is being studied. As an example, heterozygotes from the cross

h th st ss (*Aph*^{F}) ♀ x + + + + (*Aph*^{S}) ♂

were backcrossed to the female parent. Individual male progeny can then be selected that carry a chromosome recombinant at various positions, as indicated by the visible markers. These males are mated individually to virgin females homozygous for *Aph*^{F}. Six individual larvae from each mating are randomly selected for electrophoretic analysis. If the selected recombinant males still carry the *Aph*^{F} allele, originally linked to all the visible markers, all six larvae will be homozygous *Aph*^{F}. If the recombinant chromosome has picked up the *Aph*^{S} allele, some heterozygotes will be found (with a 1 in 2^6 probability of failing to find heterozygotes actually present). In MACINTYRE'S experiments, heterozygotes were never found when the *st-ss* region of the chromosome was intact. Among 114 chromosomes recombinant between *st* and *ss*, the *Aph* marker was separated from *st* only 15.8% of the time, so the locus is assigned as 15.8% of the distance from *st* to *ss*.

SCHNEIDERMAN *et al.* (1966) have reported electrophoretic variants of an alkaline phosphatase found in adults (their band 7). Using a chronological naming system, they have designated the locus so identified as *Aph*^{2} and the alleles A (fast) and B (slow). Heterozygotes have a single band corresponding to the slow form. In the F_2 generation, the segregation is three slow to one fast. This unusual behavior is most easily interpreted as indicating that B is dominant. Since virtually all other electrophoretic variants are codominant in expression, this exception invites further investigation. One clear possibility is that the locus does not code for the primary structure of the protein, but controls some modifying agent. Presence of the modifying factor (coded by B) could reasonably be dominant to absence. In any case, the locus is on chromosome II, and hence is not linked to larval *Aph*.

Mutants affecting other Aph bands have been reported, but not extensively studied. HUBBY and LEWONTIN (1966) report two electrophoretic variants of their Aph-7 in *D. pseudoobscura*. F_1 females have an intermediate band that overlaps both parental forms. F_1 males have only the form present in the female parent. Hence, this gene must be sex-linked. They also report an apparent null allele resulting in absence of their band 6. JOHNSON *et al.* (1966) found electrophoretic variants of two of the four bands they identified in *D. ananassae*. Zone B showed three variants inherited codominantly with no hybrid band. Zone D had four variants that did form hybrid bands in heterozygotes.

Development

The relative ease with which Aph is demonstrable, both in gels and histochemically, has prompted considerable work on stage and tissue specific changes in total Aph activity and Aph banding patterns. Some early work suggesting an intimate association between Aph and the DNA bands in salivary gland chromosomes (KRUGELIS, 1946) seems to be invalidated by NOVIKOFF'S (1951) observation that the classical Gomori method of staining Aph is unreliable at the intracellular level because of non-specific binding of the released phosphate—possibly to histones.

YAO (1950) used the same method, but was primarily interested in overall distribution of alkaline phosphatase during embryogenesis. He was aware of the danger of non-specific staining, and did controls in which inactivated sections were stacked alternately with active sections during staining. Absence of stain in the inactivated sections showed that there was no diffusion of sufficient scale to cause problems when dealing with the large scale aspects of the distribution. He found an abrupt appearance of Aph activity in the embryo at about the time of germ band

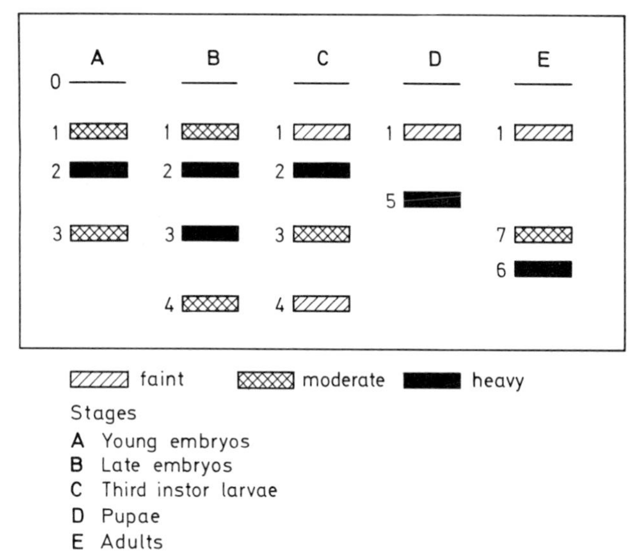

Fig. 32. Diagrammatic representation of changes in alkaline phosphatases during development. (Based on the data of SCHNEIDERMAN *et al.*, 1966)

contraction (9–10 hrs). The enzyme is localized, at first appearance, in the ventral hypodermis, and it spreads progressively from there with time. Activity subsequently disappears from most tissues prior to hatching, remaining in salivary glands and Malpighian tubules. YAO (1950) finds another dramatic increase in total Aph activity (histochemically identified) leading up to pupation, followed by a decline starting about 24 hrs after spiracle eversion.

Electrophoretic analysis of the Aph pattern in specific tissues and at various stages makes it possible to refine the developmental picture obtained by YAO. BECKMAN and JOHNSON (1964a, d) detected four different bands. One was faintly visible at all stages, a second band was visible from early stages through second instar. The prominent Aph which they studied genetically appears in the third instar and is, in turn, replaced by a characteristic pupal band.

SCHNEIDERMAN *et al.* (1966) detected a greater variety of bands and studied them in more detail. By dissection followed by electrophoresis, they were able to assign tissue localizations to most bands. Seven different bands are recognized and named in the order of appearance during development (Fig. 32). Band 1 is faintly visible in all stages, and presumably corresponds to one faint band seen at all stages by BECKMAN and JOHNSON. Bands 2, 3, and 4 appear in order during embryogenesis and all are strongly present in 18–20 hrs embryos. Band 2 is localized in larval hypodermis, and the timing of its appearance agrees with the strong stain noted by YAO at 9–10 hrs in the ventral hypodermis. The same band becomes prominent again in the third instar, and is said to correspond to the prominent larval Aph of BECKMAN and JOHNSON (1964a). This band fades during the 1st and 2nd instars, and again after pupation. Band 3 is localized in the midgut, and Band 4 (sometime resolvable into 4a and 4b) is in the midgut and hindgut. These two are present throughout larval life and disappear in prepupae about 96 hrs after hatching of the egg. Band 5 is the pupal form that replaces Band 2. It seems to be localized in the yellow body, part of the meconium derived from histolyzed larval gut cells and passed out shortly after eclosion (but see discussion below on relationship between Bands 2 and 5). Bands 6 and 7 are adult forms present in midgut and hindgut, respectively. The authors also examined some other *Drosophila* species and found patterns similar in overall plan but differing in details, such as relative mobility of the bands.

There appears to be an interesting relationship between the prominent bands in third instar larvae (Band 2) and in pupae (Band 5). The impression gained in developmental studies is that Band 2 fades exactly as Band 5 appears. Furthermore, strains with electrophoretic variants of the larval enzyme show a corresponding alteration of mobility in the pupal form (SCHNEIDERMAN *et al.*, 1966; SCHNEIDERMAN, 1967; WALLIS and FOX, 1968). SCHNEIDERMAN (1967) noted that in butanol extracted larval homogenates, the prominent larval Aph migrated in the same position as the pupal enzyme. Other Aph bands were not altered. When the hypodermis was dissected and homogenized separately from the rest of the larva, the butanol extraction did not alter the mobility of Aph. However, addition of a gut homogenate, even without butanol, did cause the modification. Boiled gut homogenate did not have the ability to alter the mobility. Bovine trypsin was also found to be active in modifying larval Aph to migrate as the pupal form. Finally, the larval gut extract was inactivated by trypsin inhibitor. All these observations suggest that pupal Aph is simply a secondary modification of larval *Aph* possibly by proteolytic action (perhaps even as an artifact of extraction). However, the reported localizations of larval Aph in hypodermis and pupal Aph in histolyzed gut cells are puzzling if this interpretation is correct.

WALLIS and FOX (1968) have examined the genetics of this relationship more carefully. On the basis of extensive (but fruitless) efforts to find recombinations separating the larval electrophoretic markers from corresponding alterations of the pupal enzyme, they conclude that the enzymes are controlled by the same locus or by linked loci no further than .0006 centimorgans apart. They also report a new variant in which pupal Aph mobilities are indistinguishable, even though the larval Aph mobilities are distinct. This is compatible with the secondary modification model if the alteration resulting in different mobilities of the larval enzyme is included in a region that is removed or masked during conversion to the pupal form.

While the exact nature of the relationship between larval Aph and pupal Aph remains somewhat open to question, this system dramatically illustrates another caution that must be observed in interpreting electrophoretic banding patterns. Distinct bands observed in a gel at various stages, or from various tissues, are not necessarily products of different genes. One can also conceive of circumstances where allelic differences in electrophoretic mobility could be due to mutations at sites other than the structural gene coding for the primary enzyme structure. If a gene coding for a modifier of some sort were mutant, that could produce an electrophoretic difference.

JOHNSON (1966a) has done some comparative studies on the developmental expression of Aph in *D. melanogaster* and *D. ananassae*. He reasons that genetically simple alterations of stage specific isozyme patterns would be most useful in analyzing the control mechanisms responsible for developmental specificity (see also the section on aldehyde oxidase). As an alternative to seeking mutations with this kind of effect, he proposed to compare related species. *D. ananassae* has a prominent late larval band that, as in *D. melanogaster*, fades and is replaced by a pupal band. Likewise, strains with larval enzyme of altered mobility show the corresponding change in the pupal enzyme. Thus, the systems appear homologous in the two species. However, the larval band reappears in adult males of *D. ananassae*. It is once again under the same genetic control as the larval band. Here, then, is a descrete difference in the pattern of expression of a particular enzyme in two species. A similar difference between strains of the same species would be most useful and exciting since it would presumably identify a gene or genes involved in differential gene expression.

Acid Phosphatases

Acid phosphatases, like alkaline phosphatases and esterases, form a loose, operationally defined group of enzymes that have been studied fairly extensively because of the ease with which these enzymes are detected in electrophoretic gels. Staining, usually at pH near 5, generally involves α-napthyl phosphate as an artificial substrate and any one of several diazonium salts to trap the released naphthol and form the stain (MACINTYRE, 1966). Electrophoresis is usually done in starch or acrylamide. A quantitative spectrophotometric assay is available using p-nitrophenyl phosphate as a substrate (MACINTYRE and DEAN, 1967).

Biochemistry

Several bands of activity are normally seen when gels are stained for acid phosphatase (Acph). However, methods vary so much that it is difficult to establish relationships from study to study. Most of the work has concentrated on one or two of the bands. MACINTYRE (1966) and MACINTYRE and DEAN (1967) have investigated Acph-1 biochemically. It is not inhibited by EDTA or citrate, but is sensitive to fluoride. A partial purification was achieved using a low pH precipitation and ammonium sulfate fractionation. Using the quantitative assay based on p-nitrophenyl phosphate, specific activity was increased 40-fold. Since Acph-1 was separated from several other phosphatases during the procedure, the actual purification was presumably higher. This purified product was inactivated at pH below 3.1 or above 10.3. The inactivation was reversible with an optimum for reversal near pH 6.5.

When partially purified extracts of strains with different electrophoretic variants are mixed, subjected to inactivating pH and reactivated at pH 6.5, an intermediate electrophoretic form (also found in heterozygotes) is produced. This suggests the enzyme has at least two similar subunits that can be dissociated at extreme pH. Molecular weight estimations based on acrylamide gel electrophoresis indicates that the subunits are, indeed, one half the size of the active enzyme (MACINTYRE, 1971a).

MACINTYRE (1971b) has used the ability of this enzyme to dissociate and reassociate, to investigate evolutionary relationships of Acph in various species of *Drosophila*. Similar work on Est-5 is described on p. 117. He argues that subunit interaction is a complex and important aspect of protein function. It will, therefore, be subject to natural selection, and is likely to be a more meaningful index of relatedness than simple similarity in electrophoretic mobility. The approach used is to dissociate enzymes XX and YY, allow them to reassociate in a mixture, and then compare the ratios of XX, XY, and YY activities to those expected on the basis of random association. Preliminary partial purification was done as described above. Separation of the reassociation products was possible by chromatography on phospho-cellulose. Alternatively, quantitation could be achieved on acrylamide gels by cutting out stained bands, eluting with glacial acetic acid and measuring the eluted stain spectrophotometrically (MACINTYRE, 1971a). In an initial study, Acph from *D. melanogaster, D. simulans* and *D. virilis* were compared. The enzymes from these three species had a pH optimum near 5.0, and all were inactivated by high pH between 10 and 11, but there were significant differences within this range. Recovery of activity upon reactivation ranged from 20% to 85%. In pairwise combinations, the mixture of *D. melanogaster* enzyme with *D. simulans* enzyme gave less hybrid enzyme than expected on the basis of random reassociation of subunits, while the *D. melanogaster* — *D. virilis* mixture gave more than expected. It should be noted that interpretation of experiments of this type is complicated by the fact that reassociation is assessed by assaying the enzyme activity in each region. Thus, the method does not distinguish between differences in the ability to form hybrid dimers and differences in the catalytic efficiency of the various dimers.

Genetics

Electrophoretic variants have been reported for several acid phosphatases in various species of *Drosophila* and one has been extensively investigated. PRAKASH *et al.* (1969) report three variants of larval *Acph-4* in *D. pseudoobscura*. The controlling gene is localized on the X-chromosome, but no further mapping is done. JOHNSON *et al.* (1966) report three alleles of *Acph-A* in *D. annanassae*. A hybrid band is formed in heterozygotes. No mapping data are reported. ESPOSITO and ULRICH (1966) also report finding electrophoretic variants, but no details are given.

More extensive work has been done by MACINTYRE (1966b) on the genetics of an acid phosphatase in both *D. melanogaster* and *D. simulans*. This is referred to as acid phosphatase-1 (Acph-1). Single flies of both species show either one band in either of two positions, or a three-banded pattern consisting of an intermediate band together with both of the single band forms. The F_1 from a cross of a true-breeding fast strain to a true-breeding slow strain produces the three-banded "hybrid" pattern. The fast form was much more common than the slow in *D. melanogaster*. Standard genetic analysis with recessive markers on chromosomes II and III shows that the electrophoretic determinant segregates with markers on III, and recombinants place the gene at 101.4 ± 0.1. In *D. simulans* the gene was mapped to $3\text{-}134.0 \pm 2.1$. Both map positions are just to the right of *claret* eye color. Homology of the enzymes in the two species is further supported by the formation of a band intermediate to the two parental types in all successful interspecies hybrids.

Acph-1 is also one of the few enzyme loci at which induced mutations have been recovered (BELL *et al.*, 1972). Mutagenesis was carried out with EMS. The screen for new mutants was dependent on the prior existence of electrophoretic variants at the locus. The essential features of the screen were as follows. Males homozygous for *Acph-1*^{A} were mutagenized and mated to females homozygous for *Acph-1*^{B}. F_1 males were mated singly to females with appropriate balancers to save any potential mutants, and were subsequently recovered and electrophoresed individually. The gels were stained for Acph. Normally, one would expect all of these males to have the typical three-banded pattern (AA, AB, BB) chracteristic of heterozygotes. A male carrying a newly induced "null" mutant would, however, produce no BB band (AB might or might not form, depending on whether or not an inactive subunit capable of dimerization was still formed). Of 2750 chromosomes tested in this manner, five confirmed *Acph-1*^{O} mutants were found. One case of conversion from *Acph-1*^{A} to *Acph-1*^{B} was also found. The authors also looked for *Est-6* "null" mutants at the same time, following a similar scheme, and found six in 2682 tested chromosomes. The frequency of induced mutations in both cases was about 0.3%.

Each of the five chromosomes carrying null mutants of *Acph-1* proved to be lethal when made homozygous. However, crosses between the different mutant stocks produced viable flies with no detectable Acph activity. Thus, the lethality must be due to mutations at other sites. Homozygous *Acph-1*^{O} stocks were ultimately obtained by crossing out most of the rest of the chromosome and retaining only the region around *Acph-1*. This procedure is tedious, and the authors point out that if one desires to obtain homozygous mutants induced by EMS at a specific locus, one ought to choose loci that are easily isolated genetically or use low doses of EMS.

Once some null mutants were obtained, a method of screening without electrophoresis was possible. Using the induced null alleles as controls, conditions were found for a spot test for Acph activity specific for Acph-1. The screen was now done by mutagenizing *Acph-1*^{A} or *Acph-1*^{B} males and mating them to females that were *TM3, Sb Ser/e Acph-1*^{O} (homozygous lethal). Phenotypically wild-type F_1 males (*e Acph-1*^{O}/mutagenized *Acph-1*^{A} or *Acph-1*^{B}) were mated singly to save the chromosomes, and then recovered and subjected to the spot test for *Acph*. Newly induced null alleles were indicated when the F_1 male had no activity. Eleven new *Acph-1*^{O} alleles were obtained using this screening method. The frequency was one mutant in every 400–500 chromosomes tested.

Development

Only a very sketchy picture of developmental changes of Acph is available. ESPOSITO and ULRICH (1966) report that various bands appear and disappear as a function of stage, but no specifics are given. PASTEUR and KASTRITSIS (1971) report some degree of tissue specificity among 4 Acph bands they detected in whole flies. For example, salivary gland contains only bands II and III, while fat body has no activity. They also noted some developmental changes. Band II increases in intensity in the period following spiracle eversion, while band III disappears within 12 hrs after this event. Clearly, there is room for much more work on the developmental histories of the various acid phosphatases.

Miscellaneous Enzymes

Carbohydrate Metabolism

Sucrase and Trehalase (EC 3.2.1.26), (EC 3.2.1.28)

MARZLUF (1969) and HUBER and LEFEBVRE (1971) have described enzymes which can hydrolyze the dissaccharides sucrose and trehalose, respectively. The hydrolysis of various dissaccharides and trissaccharides by crude extracts has shown that enzymatic activity is specific for the glucosidic linkage. The rate of hydrolysis of sucrose is about four times that of trehalose. Melezitose, turnanose and maltose are cleaved at slower rates. Crude extracts do not hydrolyze methylglucoside, methylmannoside, lactose or raffinose (MARZLUF, 1969). Sucrase and trehalase activities represent hydrolysis by different enzymes. This has been shown by demonstrating several differences in the two activities. These include differences in heat inactivation kinetics, low pH inactivation, pH optima, and sensitivity to urea. In addition, these enzymes can be separated from one another by means of DEAE ion exchange chromatography and acrylamide gel electrophoresis.

Biochemistry of Sucrase

Biochemical properties of sucrase in crude extracts have been studied by MARZLUF (1969), and in a partially purified preparation by HUBER and LEFEBVRE (1971). MARZLUF reported a substrate K_m for sucrase of 33 mM, while a K_m of 43 mM was reported by HUBER and LEFEBVRE. MARZLUF observed a pH optimum of 6.5, while HUBER and LEFEBVRE report a pH optimum of 5.6. Heat inactivation studies using crude extracts showed a two-phase curve with approximately 10% of the activity stable at 46 °C (MARZLUF, 1969). The partially purified enzyme shows only a single form in heat inactivation studies performed at 45 °C, (HUBER and LEFEBVRE, 1971). The enzyme is strongly inhibited by TRIS buffers in crude extracts, or after purification. HUBER and LEFEBVRE (1971) report the enzyme is stimulated by 2-mercaptoethanol. MARZLUF (1969) found the enzyme eluted in the void volume on Sephadex G-200 gel filtration. However, HUBER, and LEFEBVRE (1971) estimated the M.W. of sucrase on calibrated G-200 columns to be less than 100000, and have observed that the G-200 elution position of sucrase changes with the degree of purification. With increasing purification a lower M.W. is indicated. HUBER and LEFEBVRE report that, in general, sucrase activity is not very stable. They have obtained partially purified enzyme by combining ammonium sulfate precipitation, ion exchange chromatography and preparative disc-gel electrophoresis. These preparations are 87-fold purified, and represent 2.4% of the starting sucrase activity. MARZLUF has shown that sucrase is distributed in soluble and particulate fractions. The particulate enzyme, sedimentable at 130000 × g, can be solubilized by treatment with the non-ionic detergent, triton X100. HUBER and LEFEBVRE have suggested that some of the differences between their results, using partially purified enzyme and MARZLUF'S data using crude extracts, can be accounted for if the particulate and soluble forms have different properties. MARZLUF (1969) has studied the soluble, particulate and solubilized forms of sucrase with respect to substrate K_m and some stability properties, and finds no significant differences. He also noted that sucrase has five or six isozymes, visualized as activity bands in acrylamide gels. There are preliminary indications that some of these isozymes have different properties with respect to stability and behavior during purification. Since the partially purified preparations of HUBER and LEFEBVRE were not examined for isozymes, it is not possible at this time fully to account for the differences in the properties of sucrase in the crude extracts reported by MARZLUF, and in partially purified preparation of HUBER and LEFEBVRE. They could be the result of differences between soluble and particulate enzyme, both of which are present in MARZLUF'S preparations, or these differences may be the result of HUBER'S and LEFEBVRE'S selectively purifying one of the isozymic forms of sucrase.

Biochemistry of Trehalase

The properties of trehalase have also been reported by MARZLUF (1969) and HUBER and LEFEBVRE (1971). In general these studies led to similar conclusions. Trehalase is distributed between the soluble and particulate fractions of the cell. The particulate fraction is obtained by centrifugation at 130000 × g of a 12000 × g supernatant. The pH optimum is 5.6. The K_m for trehalose is 0.136 mM. The enzyme is inhibited by sucrose. It is also inhibited by TRIS but to a lesser degree than sucrase. As judged by heat inactivation kinetics, fractionation on DEAE ion exchange chromatography and acrylamide gel electrophoresis, trehalase apparently exists in only one form. Purification has been accomplished by HUBER and LEFEBVRE. The preparation resulting from ammonium sulfate fractionation, ion exchange chromatography and preparative gel electrophoresis is 1086-fold purified with respect to trehalase activity. The enzyme is obtained in a yield of 18.4% of the starting activity. The preparation obtained by HUBER and LEFEBVRE shows only one protein band upon analytical acrylamide gel electrophoresis. The molecular weight of purified trehalase is somewhat less than 100000, as judged by gel filtration on G-200.

Developmental Biology of Sucrase and Trehalase

Constant and relatively low levels of both enzymes are observed in larval and early pupal life. Just prior to emergence the specific activities begin to rise and reach a maximum value during the first day of adult life which is seven or eight times the larval level. This rise in activity is independent of different sugars in the diet, so there does not appear to be a substrate induction phenomenon involved in the rise in activity (MARZLUF 1969).

Genetics of Sucrase and Trehalase

The levels of sucrase and trehalase vary in a number of different strains which were analyzed by MARZLUF (1969). Furthermore, the two enzymes appear to be independent of one another, resulting in different ratios of sucrase to trehalase activity in different strains. Oregon R shows a ratio of sucrase to trehalase of about 1.3, while the ratio in Basc strain is 0.5, and in a strain carrying *cn, bw; e,* it is 5.0. MARZLUF (1969) reports some strain variation in the isozyme pattern of sucrase. Electrophoretic variants have not been identified as yet, and further genetic characterization of this enzyme is presently lacking.

Hexokinase (EC 2.7.1.1)

Hexokinase in *Drosophila* has been the subject of a few papers dealing primarily with zymogram patterns on electrophoretic gels. The enzyme, which catalyzes the formation of glucose-6-phosphate from glucose and ATP, can be stained by adding glucose-6-phosphate dehydrogenase to couple the product to the standard tetrazolium reduction system used in staining various dehydrogenases. Activity toward other hexose sugars may be detected if phosphohexose isomerase is added to the staining mixture.

Biochemistry of Hexokinase

Typically, there are a number of distinct bands of hexokinase activity separable by electrophoresis. MURRAY and BALL (1967) report 5 zones of activity in *D. melanogaster*, KNUTSEN *et al.* (1969) report 4 zones, some with multiple bands, in *D. robusta* and JELNES (1971) finds 3 bands in *D. melanogaster*. MADHAVAN *et al.* (1972) find a total of four bands in *D. melanogaster*. Thus far, biochemical characterization has not proceeded beyond attempts to determine substrate specificity by observing relative staining intensities in gels. Of the four zones in *D. robusta,* 3 and 4 are equally active toward glucose and fructose, while zone 1 has considerably less activity toward fructose (KNUTSEN *et al.*, 1969). However, zone 1 isozyme obtained from muscle does use fructose as a substrate, while zone 1 isozyme from digestive tract does not. This suggests the existence of two distinct enzymes with similar electrophoretic mobility. These authors also report that *D. melanogaster, D. willistoni,* and *D. prosaltans* have basically similar patterns (4 zones), differing only in absolute migration. All three of the zones observed by JELNES (1971) in *D. melanogaster* are active with glucose as substrate. However, only the fastest zone uses mannose as a substrate, and the slowest is more active with fructose. On this basis, the three zones, from fastest to slowest, have been designated Mk (mannokinase), Gk (glucokinase) and Fk (fructokinase). These designations seem premature in the absence of more detailed enzymological characterization of the three isozymes. The basis of the various number of zones (3, 4, or 5) reported for *D. melanogaster* is not clear, and dramatizes the difficulty of relating the nomenclature of one author to that of another.

Genetics of Hexokinase

KNUTSEN *et al.* (1969) found apparent allelic variants for some of the isozymes in *D. robusta*. Region 3 is genetically the simplest. There are four bands in the zone, 2 major and 2 minor, that are controlled by two codominant alleles. Presumed homozygotes contain 1 major and 1 minor band, and the major and minor band are shifted coordinately in the two types of homozygotes. Heterozygotes contain all four parental bands but no new hybrid bands.

Region 1 contains 2 bands also thought to be controlled by a pair of codominant alleles. However, the region 1 isozyme from the digestive tract forms bands at positions 1 and 2, but with no apparent genetic variations, while muscle enzyme from the same region does show genetic variation. This supports the evidence obtained in the substrate specificity studies in suggesting different isozymes in the two tissues.

Region 2 contains bands 3–8. The genetics is not worked out, but there is some indication that all six bands are shifted coordinately. No variants were reported in the single band in region 4.

JELNES (1971) found electrophoretic variants for the slow (fructokinase?) band in *D. melanogaster.* Again, they appear to be controlled by a pair of codominant alleles, and the heterozygote contained the parental bands but no hybrid form. The electrophoretic markers were found to segregate with 2nd chromosome markers, and the gene controlling the difference was mapped to 2-73.5 on the basis of 108 chromosomes recombinant between *cn* (2-57.5) and *c* (2-75.5). MADHAVAN *et al.* (1972) find a comparable map position (2-73 ±) for their *Hex-3.*

Developmental Biology of Hexokinase

The same four papers provide considerable information on the localization of the various isozymes and changes in pattern during development. Some of the most interesting observations concern sex specificity of some of the bands.

The slowest of the 5 bands reported by MURRAY and BALL (1967) ("Zone A" or *Hex-1* in their nomenclature) is found only in males and not in adult females. Attached-X females (XX/Y) do not have this band and XO males do, so it is not Y-linked, but rather, sex-limited. The band is fond in larvae and pupae of both sexes, but disappears in females within two days after eclosion. Females carrying the *transformer* gene *(tra/tra)*, which have a male-like phenotype even though genetically female, also show *Hex-1* activity.

By electrophoretic analysis of extracts prepared from dissected tissues, *Hex-1* was localized in the male accessory glands. In the course of this dissection work, a distinct band not previously noted in whole fly extracts was found in testes and designated Hex-t. This enzyme, too, was present in XO males and *tra/tra* females, and hence is sex-limited, not sex-linked. MADHAVAN *et al.* (1972) also report a male specific hexokinase localized in testes.

Although Hex-1 is present in 3rd instar larvae and pupae, the band was never as intense as in adult males, and no tissue localization could be demonstrated. Hex-t is first detectable in testes from early pupae. It is worth noting that these authors report that *D. pseudoobscura* has a different hexokinase pattern and no sex specific enzymes. MURRAY and BALL also noted stage specific changes in other bands. For example, Zone B gets stronger in pupae, while Zone D disappears.

MADHAVAN *et al.* (1972) examined both stage specificity and tissue specificity of the three bands in addition to Hex-t. Hex-1 (and Hex-t) were detectable only in adults. Hex-2 is present at all stages, while Hex-3 is apparently absent in freshly laid eggs but is found at all later stages examined. Tissue localization, done by electrophoresing extracts of dissected tissues, revealed several interesting features. Hex-1, the adult specific band, is faintly detectable in larval imaginal discs. Hex-2 is widely distributed in larval tissue, while Hex-3 is concentrated in fat body. Hex-t could not be detected in larval testes. In adults, Hex-2 and Hex-3 are widely distributed (although the largest amount of Hex-3 is found in mid-gut). Hex-1 is localized in thoracic muscles and accessory glands. The fact that MURRAY and BALL (1967) found their Hex-1 to be male specific and localized in accessory glands, and found a total of 5 bands as compared to four reported by MADHAVAN *et al.* (1972), suggests that the agar gel method used by the latter group fails to resolve two bands that were resolved in the earlier work. On the other hand, the stage specificities reported by the two groups contain some discrepancies if this interpretation is followed.

KNUTSEN *et al.* (1969) have also studied the tissue distribution of the hexokinase isozymes in *D. robusta*. Isozymes in region 2 are widely distributed in most or all tissues. Region 1 isozymes are found in digestive tissue and muscle, but evidence has been mentioned above that suggests control by different genetic loci. Region 3 hexokinases are found in testis, muscle and digestive tissue. Region 4 is male specific and found only in testis. These authors also noted changing patterns with time that suggest stage specific expression of some of the hexokinases.

The work done thus far on hexokinases indicates it to be a notably interesting system. There is need for considerably more biochemical and genetic work, and opportunity for additional, interesting, developmental analysis.

β-Glucuronidase (EC 3.2.1.31)

The levels of β-Glucuronidase found during embryogenesis have been studied by BILLET and COUNCE (1957). They have found appreciable levels of this enzyme in stages from cleavage through hatching. The level of activity does not vary significantly during this period of development. Later stages have not been analyzed.

Aldolase (EC 4.1.2.7)

Aldolase is an important enzyme in the glycolytic pathway that cleaves fructose-1, 6-diphosphate into two triose phosphates. Electrophoretic detection of the enzyme is possible by adding α-glycerophosphate dehydrogenase and triose isomerase to couple the products to a tetrazolium reduction system (BRENNER-HOLZACH and LEUTHARDT, 1967). It has been the subject of biochemical investigations by BRENNER-HOLZACH and LEUTHARDT (1967, 1968, 1969), but has yet to be investigated genetically or developmentally. A purification procedure employing a mild heat treatment, ammonium sulfate precipitation, chromatography on DEAE-cellulose, and gel filtration on Sephadex G-200 yields a product that is 70–90 fold purified, and that can be crystallized. Activity of the enzyme is not dependent on any tested metal ion, and is not sensitive to EDTA. It has a broad pH optimum from 6.8 to 8.0. The molecular weight, estimated from sedimentation data and gel filtration, is about 160000. The purified preparation yields a single band of protein after either disc electrophoresis or cellulose acetate strip electrophoresis. Disc electrophoresis in 8M urea yields two bands, and a sharp decrease in the sedimentation constant is also noted in urea. These data suggest an enzyme composed of subunits of at least two types. The K_m was measured with both fructose-1, 6-diphosphate and fructose-1-phosphate. The values obtained are similar to those known for rabbit muscle aldolase, but the ratio of V_{max} (FDP/F-1-P) is much lower.

An antibody prepared against the purified enzyme gives a single precipitation band when reacted against either the purified enzyme or a crude extract. The antibody does not cross-react with either rabbit muscle or rabbit brain aldolase. An amino acid analysis also shows significant differences from the rabbit enzymes. The *Drosophila* enzyme is, however, capable of forming apparent hybrids with rabbit brain aldolase after dissociation at pH 2. Three new electrophoretic bands are found upon reassociation even though the mobilities of the two parent enzymes are similar, and neither parent enzyme forms extra bands when dissociated and reassociated alone.

No electrophoretic variants were reported in this work, and GILLESPIE and KOJIMA (1968) also failed to find any variants in a fairly extensive survey of two wild populations of *D. ananassae*.

Phosphoglucomutase (EC 2.7.5.1)

Relatively little work has been done on phosphoglucomutase in *Drosophila*. The enzyme has been studied exclusively in starch gel zymograms using glucose-1-phosphate as substrate and coupling the product, glucose-6-phosphate, to a tetrazolium reduction system through glucose-6-phosphate dehydrogenase. HJORTH (1970) and TRIPPA *et al.* (1970) both report two distinct bands, with single flies containing either one or both. The two bands are apparently controlled by a pair of codominant alleles, and no hybrid enzyme is formed in heterozygotes. HJORTH (1970) located the *Phosphoglucomutase (Pgm)* locus on the 3rd chromosome by showing that it segregated with *th*. He mapped it to 3-43.4 on the basis of 30 chromosomes recombinant between *th* (3-43.2) and *st* (3-44.0).

A natural population studied by TRIPPA *et al.* (1970) was polymorphic for the two variants, but all of ten laboratory populations contained only one or the other allele. All developmental stages have the same phosphoglucomutase band. TRIPPA *et al.* (1970) found a map position of 3-43.6 based on 54 recombinants between *Gl* (3-41.4) and *st* (3-44.0). This is in very good agreement with HJORTH'S estimate. HJORTH uses the notation Pgm^1 and Pgm^2, with the corresponding alleles in TRIPPA *et al.* being Pgm^A and Pgm^B.

Fumarate Hydratase (EC 4.2.1.2)

Fumarate hydratase has recently been assayed in agar gels after electrophoresis by MADHAVAN and URSPRUNG (1973). In most strains a single zone of activity is found. One strain was found to be polymorphic, containing two electrophoretic forms. The gene responsible for the electrophoretic variant was mapped to 1-19.9.

The developmental profile of fumarate hydratase has been reported by WHITNEY and LUCCHESI (1972). Activity per organism rises during larval growth, declines following pupation and rises to maximal levels following emergence. The levels found in adults are about four-fold higher than late larvae. Females have a little less than twice the fumarate hydratase content of males. However, when expressed on a specific activity basis, adult males have a somewhat higher level. This is presumably due to fumarate hydratase deficient eggs in the female.

Nucleic Acid Metabolism

Deoxyribonucleases

In a series of investigations BOYD has identified several enzymes in *Drosophila* that have deoxyribonuclease activity, and has proceeded to characterize their biochemical and developmental properties. The identification of multiple deoxyribonucleases was made possible by means of an assay, developed by BOYD and MITCHELL (1965), which allows the visualization of zones of deoxyribonuclease activity in polyacrylamide gels after electrophoresis. This assay is based on the inclusion of DNA in the acrylamide solution during polymerization, and electrophoresing the enzyme sample into this DNA-gel. The gel is then incubated under conditions appropriate for enzymatic catalysis. After incubation the digestion products are removed, and the gel is stained with a DNA stain. The gels can be observed visually and photographed or scanned with a densitometer. Activity bands appear as unstained areas in the gel where the DNA has been digested. BOYD (1969) successfully applied this procedure to the analysis of deoxyribonuclease present during development.

Developmental Biology of DNase

By estimating activity from the number and intensity of enzyme bands in gels, it was concluded that third instar larvae have a particularly high activity. These high

Table 4. Tissue Distribution of DNase Isozymes

Tissue	Bands in larvae	Bands in pupae	Possible number common to both stages[a]
Salivary glands	1	6	1
Malpighian tubules	1	4	1
Mid-intestine	3	6	0
Fat body	8	7	4
Hemolymph	4	6	4
Blood cells	3	4	3

[a] These numbers were arrived at by our analysis of the data of BOYD and BOYD. It is at best an approximation, and is intended only as a depiction of the large changes which occur.

levels begin to fall as pupation is approached. In late third instar larvae (115 hrs at 25° C), quantitative and qualitative changes are evident. There is a drop in total activity, the disappearance of some bands and the appearance of new bands. Pupae have relatively lower activities, and flies several days after eclosion again reach higher levels. These results of BOYD (1969) agree relatively well with those of MUHAMMED *et al.* (1967), who measured total deoxyribonuclease activity throughout development using the reduction in the transformation capacity of treated bacterial DNA as an assay for deoxyribonuclease. During development, highest levels are seen in third instar larvae, lower levels in pupae and intermediate levels in adults. BOYD (1969) and BOYD and BOYD (1970) have extended the developmental studies to measure changes in the number and types of deoxyribonucleases in various organs and tissues as development proceeds. BOYD (1969) has concluded that there is a minimum of seven different deoxyribonucleases which are active at pH 4.0. This conclusion was reached by noting preference for native or denatured DNA as substrate, the effect of Mg^{++}, or the effect of EDTA. The pH 4 active deoxyribonucleases are found predominantly in the intestine and hemolymph. More extensive analyses on organs from third instar larvae and pupae have been performed by BOYD and BOYD (1970). The transition period between the larval stage and pupal stage is a period in which marked changes in deoxyribonucleases might be expected. The larvae are ceasing to feed, and those enzymes primarily involved in digestion might decline. The prepupa is preparing for metamorphosis and entering a period of extensive histolysis, and the appearance of new enzymes for DNA degradation might be expected to accompany this process. BOYD and BOYD (1970) separated the deoxyribonucleases from several organs by electrophoresis, and tested the activities at pH 4, 7, and 9, the effect of EDTA, Mg^{++} ions and the preference for native or denatured DNA as substrate. These enzymatic properties, in addition to their relative electrophoretic mobilities, allowed different deoxyribonucleases to be distinguished and an estimate made of the number in different organs and tissues.

The data in Table 4 only give a general survey of the changes occurring in each tissue at these stages. Identity of particular enzymes is difficult to ascertain and as pointed out by BOYD (1969), the absence of a particular band in a gel cannot be considered as indicating an absolute difference, especially when other active bands are present in a similar region of the gel. Nonetheless, it is quite clear that major changes in deoxyribonucleases are occurring in several tissues as the animals enter pupation. The salivary gland, Malphighian tubules and mid-intestine show the most dramatic shifts, while fat body, hemolymph and blood cells show a somewhat smaller but significant shift in total pattern.

As a check on the validity of this method of comparison, BOYD and BOYD (1970) have mixed extracts of more than one tissue, and found that the gel patterns are additive when compared to the tissues run separately. In addition, several enzyme bands were eluted from the gels and re-electrophoresed. In these experiments it was seen that the enzymes have the same relative mobilities in the second run as they did initially in crude extracts.

BOYD has grouped together deoxyribonucleases with similar properties for the purpose of making functional comparisons. Nucleases active at higher pH show a marked tissue specific distribution, and show major changes at the onset of pupation. The deoxyribonucleases of this class present in the hemolymph and intestine show a sharp decline at this stage, and BOYD and BOYD (1970) suggest that the intestinal enzyme might be involved in digestion, the decline being consistent with cessation of feeding. The deoxyribonucleases active at acid pH do not show as pronounced tissue specific differences. One class, appearing simultaneously in several tissues which will undergo histolysis, or at least major changes during metamorphosis, have properties that are consistent with the interpretation of their being lysosomal deoxyribonucleases.

Biochemistry of DNase

BOYD (1970a and 1970b) has proceeded to more carefully characterize some of the deoxyribonucleases present in salivary glands. The salivary glands are a particularly attractive tissue in which to study enzyme changes. Having been the object of extensive cytological investigations, it offers the possibility of associating a particular enzyme with a particular puff. This association might enable one to study the initiation of enzyme synthesis, using cytological and genetic approaches as well as biochemical approaches.

BOYD (1970b) has characterized an acid active deoxyribonuclease that is found in larval salivary glands as well as most other tissues. Enzyme preparations were obtained from isolated salivary glands prepared in bulk by the method of BOYD *et al.* (1968), and assayed by following the degradation of radioactive *E. coli* DNA. The pH optimum was found to vary with Mg^{++} ion concentration and to range between 4.7 and 5.5. However, Mg^{++} is not a requirement for the enzyme, since EDTA has no effect on its activity. Activity is strongly affected by high NaCl concentration, 0.2 M giving complete inhibition, and by a variety of divalent cations. Listed in order of effectiveness of inhibition these include Mn^{++}, Cd^{++}, Zn^{++}, Co^{++}, and Cu^{++}. The molecular weight has been estimated from sedimentation in sucrose gradients to be 41000. The mode of enzymatic activity is apparently that of an exonuclease, and it is more active on native than on denatured DNA. The limits of degradation appear to be about 50% of the substrate DNA.

This acid active nuclease is apparently a single enzyme. BOYD (1970b) reached this conclusion from evidence that several methods of fractionation, electrophoresis in acrylamide gels, sedimentation in sucrose gradients and pH optima studies all show one single, symmetrical peak of activity. In all of its properties this salivary gland enzyme seems identical to an acid nuclease found in the larval hemolymph.

BOYD (1970a) has characterized a second deoxyribonuclease found in the salivary gland. This enzyme has a pH optimum of 8.6, is activated by Mg^{++} which is required for activity, and is completely inhibited by 2×10^{-3} M EDTA. Other divalent cations Co^{++}, Mn^{++}, and Zn^{++} activate to a lesser degree. Cu^{++}, Cd^{++}, Ba^{++}, and Ca^{++} are inhibitory. The enzyme is inhibited by RNA and NaCl, 10% of control activity remains in 0.27 M NaCl. Li^{+} or Cs^{+} has little effect on activity. This enzyme is an endonuclease that apparently produces closely aligned double strand scissions. It is more active towards native DNA. Its molecular weight has been estimated by sucrose gradient sedimentation analysis to be 104000.

One of the most interesting aspects of this alkaline active nuclease is its developmental history. It is not detectable in late third instar larvae, but appears shortly after puparium formation. By 10 hrs after puparium formation it is no longer detectable. This very precise, abrupt regulation suggests that this enzyme is produced for some specific function at this stage in the developmental history of the salivary gland. This is a period which follows marked alterations in the puffing activity of the polytene chromosomes. URSPRUNG *et al.* (1968), BOYD (1970a) and many other investigators, e.g. ASHBURNER (1970), have pointed out that these chromosomes provide a number of advantages for studying gene regulation during development. Workers involved in these studies have had a desire to associate a specific, biochemically characterized gene product with a specific puffing site. The combination of its developmental history, tissue specificity and biochemical accessibility make the alkaline active deoxyribonuclease a likely system in which a gene product-puff association is likely to be made in the near future.

Guanine Deaminase, Inosine Phosphorylase, Adenosine Deaminase

HODGE and GLASSMAN (1967) have assayed several enzymes which are involved in purine catabolism in *Drosophila*. The levels of guanine deaminase, inosine phosphorylase and adenosine deaminase have been measured in extracts of third instar larvae of various strains and mutants. Different levels of activity are characteristic of specific strains and mutants. The mutant *brown (bw)* has 50% of wild type levels of guanine deaminase. However, it has not been possible to associate any one of these enzymes with a specific locus. UDA *et al.* (1969) have also detected an activity that deaminates adenosine or adenosine 2′,3′ cyclic phosphate. This activity has been purified 30-fold by means of ammonium sulfate fractionation and gel filtration chromatography. The ratio of activity using adenosine 2′,3′ cyclic phosphate as substrate in comparison to adenosine is 0.13.

Thymidylate Synthetase

The enzyme thymidylate synthetase has been the object of a brief report by MAGDON (1969). He noted a marked inhibition, up to 50%, of thymidylate synthetase activity in extracts prepared from irradiated flies. MAGDON suggests that this is the explanation for the elevation of the mutation rate which is seen if 5-bromodeoxyuridine is applied following irradiation. The radiation would inhibit thymidylate synthetase, and thereby prevent the metabolism of bromodeoxyuridine. This would allow it to be more readily incorporated during DNA synthesis and subsequently to cause mispairing.

Aminoacyl-tRNA Binding Enzyme

In recent years the enzymatic machinery required for protein biosynthesis has been studied in a variety of prokaryotic and eukaryotic systems. This work has recently been reviewed by LUCAS-LENARD and LIPMANN (1971). Investigations by STAFFORD and co-workers have begun to extend these observations to *Drosophila*. PELLEY and STAFFORD (1970a,b) have described and partially purified an enzyme which catalyses the codon specific binding of amino acyl-tRNA to ribosomes. Specifically, the binding of phenylalanine-tRNA to *Drosophila* ribosomes promoted by poly U and the *Drosophila* binding enzyme was measured. Partial purification of this enzyme from the 230000 × g supernatant was effected by sequential ammonium sulfate fractionation, calcium phosphate gel adsorption, Sephadex G-200 gel filtration chromatography and sedimentation through sucrose gradients. The resulting enzyme preparation represented a 24.8-fold purification, and a recovery of 3.9% of the activity from the initial 230000 × g supernatant. The molecular weight of this binding enzyme, estimated from cicalibrated G-200 gel filtration columns, is on the order of 300000. However, the molecular weight estimated from sucrose gradient velocity sedimentation is 193000. This discrepancy may indicate a high degree of asymmetry in the molecule.

PELLEY and STAFFORD (1970b) have studied the requirements of the enzyme in the binding reaction. Aminoacyl-tRNA binding is proportional to the amount of added ribosomes, poly U, aminoacyl-tRNA and binding enzyme. The reaction requires guanosine triphosphate (GTP) or 5′-guanylmethylenediphosphate. It has a magnesium ion optimum between 5 and 7 mM. Potassium ions are somewhat inhibitory. It does not require reduced thiols. An enzymatic factor which is involved in peptide polymerization has also been described by PELLEY and STAFFORD (1970b). This factor is apparently released from ribosomes by washing with 0.5 M KC1. The polymerization factor can be recovered in the 60–70% saturated $(NH_4)_2SO_4$ fraction of the 230000 × g supernatant.

Aminoacyl-tRNA Synthetases

CHRISTOPHER *et al.* (1971) have described some of the biochemical properties of phenylalanine-tRNA synthetase from *Drosophila* larvae. They have also reported the measurement of leucine-tRNA synthetase and isoleucine tRNA synthetase activities. Phenylalanine-tRNA synthetase shows a broad pH optimum from pH 6.5 to 8.5. Optimum rates of activity are observed at 37° C. Maximal activity is observed when the ratio of Mg^{++} to ATP in the reaction is 25:1, specifically 50 mM Mg^{++}. Gel filtration on Sephadex G-200 using a series of markers has indicated that the molecular weight of phenylalanine-tRNA synthetase is close to 200000. Sucrose gradient sedimentation analysis indicates a molecular weight of 181000 or 159000, depending on the marker comparison.

Leucine-tRNA synthetase shows somewhat different properties. It has a Mg^{++} optimum of 10 mM and Mg^{++} to ATP ratio of 5. Its molecular weight estimated by gel filtration chromatography is on the order of 75000.

Isoleucine tRNA synthetase was also assayed in these studies. The only reported property is that it shows a Mg^{++} optimum of 10 mM and Mg^{++} to ATP ratio of 5.

Ribonucleic Acid Polymerase

PHILLIPS and FORREST (1973) have recently reported the first observations on DNA dependent RNA polymerase from *Drosophila*. The enzyme has been assayed in isolated nuclei, extracted and partially purified. When assayed in isolated nuclei from embryos the enzyme has a Mg^{++} ion optimum of 1.5 mM, a Mn^{++} ion optimum of 0.75 mM, and an $(NH_4)_2SO_4$ concentration optimum of 0.1 M. The enzyme is inhibited 40 to 60% by α-amanitin.

The enzyme has been solubilized from nuclei. DEAE cellulose chromatography of the extracted enzyme shows two peaks of activity. These have been referred to as Form I and Form II. Both are dependent on Mg^{++} and Mn^{++} ions, and are inhibited by high ammonium sulfate concentration. Only Form II is sensitive to α-amanitin.

Phosophocellulose chromatography of RNA polymerase extracted from nuclei yields a single peak of activity which contains both Form I and Form II activities. Chromatography of the phosphocellulase peak on DEAE-cellulose gives four fractions of RNA polymerase activity. Two of these are α-amanitin-sensitive and two are α-amanitin-resistant, and have been identified as Forms IIa, IIb, 1a and 1b respectively. Therefore it appears that RNA polymerase from *Drosophila* embryos exist as two or possibly four species.

Other Enzymes

Cytochrome Oxidase (EC 1.9.3.1)

Developmental Biology of Cytochrome Oxidase

Cytochrome c Oxidase is an enzyme universally found in mitochondria. FARNSWORTH (1964) has verified that this enzyme is also found in the mitochondrial fraction of *Drosophila* extracts. The initial studies on cytochrome oxidase of *Drosophila* were carried out by BODENSTEIN and SACKTOR (1950). These workers conducted a study of the activity levels which are found during metamorphosis. Activity levels begin to fall at the onset of pupation, and reach minimum two days after pupation. Cytochrome oxidase activity then increases to reach a maximum level two days after emergence. This approximately U-shaped developmental profile is characteristic for several insects (BODENSTEIN and SACKTOR, 1952). It has been confirmed in *Drosophila* by the more recent observations of WARD and BIRD (1962). FARNSWORTH (1964) has examined the levels of cytochrome oxidase found in extracts and in mitochondrial fractions obtained from developing larvae. She has shown that the activity per mg mitochondrial protein remains constant throughout larval development, except for a noticeable drop at the second molt. The specific activity also drops at the onset of pupation. These data do not indicate directly the total amount of activity per animal. One might expect that activity levels of an enzyme such as cytochrome oxidase, which plays a central role in oxidative metabolism, would parallel the growth of the animal. This assumption predicts an increase in activity per animal and an approximately constant specific activity per mg protein during larval growth. These observations, showing variation in activity per unit mitochondrial protein, could indicate the existence of mitochondrial-like particles devoid of cytochrome oxidase, and may provide an experimental system in which the development of enzyme activities in mitochondria could be studied.

Genetics of Cytochrome Oxidase

A search for genetic factors which influence the cytochrome oxidase system has been pursued by several investigators. WARD and BIRD (1962) compared the levels of cytochrome oxidase found in the Oslo strain and the Iso Amherst strain of *Drosophila melanogaster*. Both strains exhibit comparable developmental profiles through pupation and the first eight days following emergence. However, the Oslo strain has higher levels of cytochrome oxidase expressed on a mg live weight basis. Subsequently, WARD and BIRD (1963) constructed lines with chromosomes interchanged between the two strains. The higher levels of cytochrome oxidase activity were consistently found in lines carrying the second chromosome from the Oslo strain. Genetic characterization of the second chromosome factor(s) has not been pursued further.

FARNSWORTH has conducted a series of investigations aimed at elucidating the molecular basis of the *Minute* phenotype. The *Minute* mutants are characterized as being a class of dominant mutants which are lethal when homozygous and map at a large number of loci. They all produce a generalized weakness of the animal, delay in development and often secondary defects such as a small body size, missing bristles and aristae and abnormal wings.

She began her investigations on the $M(2)1^2$ by comparing the developmental aspects of cytochrome oxidase in this mutant with wild-type. The profile of cytochrome oxidase activity during larval life of the *Minute* is similar to wild-type, except that the rise in activity which follows the second molt is delayed. Larval organs at this time (60–65 hrs) such as body wall, gut and Malpighian tubules showed substantially lower total activity in the *Minute*. However, activity in imaginal discs and brain were comparable to wild-type. In older animals the cytochrome oxidase activities in *Minute* $M(2)1^2$ and wild-type were found to be comparable. In later experiments FARNSWORTH (1965a) examined the specific activities of cytochrome oxidase in mitochondria isolated from 10 different *Minutes*, and could show no correlation of the *Minute* phenotype with specific alterations in cytochrome oxidase. FARNSWORTH proposed that *Minutes* represent alterations, at different sites, in some cellular process common to most cells. FARNSWORTH suggested that the general process altered in these mutants is oxidative phosphorylation, and she conducted experimental analyses to test this hypothesis (FARNSWORTH, 1965b). Mitochondria were isolated and tested for phosphorylating ability and oxygen utilization. *Minutes* were observed often to have a lower P/O ratio. However, these results are not clear enough to establish a causal connection between the biochemical lesion and the mutant phenotype.

Another basis for the *Minute* phentoype has been suggested by RITOSSA *et al.* (1966). These authors suggested that *Minutes* represent alteration in the genes coding for transfer RNA. This hypothesis would also be consistent with the common aspects of the *Minute* phenotype, and suggests they all suffer from defects in their protein synthesizing capacity due to a missing tRNA. There may be on the order of 55 different *Minutes,* which is a number compatible with the expected number of tRNA loci. In addition, the *Minutes* have genetic properties which indicate many of them may be deletions. RITOSSA *et al.* have estimated that each tRNA gene has about 12.5-fold redundancy. If this is so, then one might expect only a deletion of all or most of the copies would produce a major alteration in phenotype. Recently, STEFFENSEN and WIMBER (1972), using the technique of *in situ* DNA-RNA hybridization to salivary chromosome preparations, have begun to localize the sites where tRNA will hybridize. Further approaches along this line should soon reveal whether there is a coincidence of tRNA loci and the cytogenetic position of the *Minute* mutants.

Catalase (EC 1.11.1.6)

Catalase activity has recently been measured in adult flies by SAMIS *et al.* (1972). The enzyme shows a pH optimum of 7.5. It is inhibited by increasing concentrations of H_2O_2, its substrate. The assay for catalase activity is not appreciably affected by ionic strength.

Catalase specific activity is highest in the abdomen, as compared to lower levels found in the thorax and head. Male abdomens have a significantly higher specific activity than female abdomens, while the head and thoracic levels are similar in the two sexes. This observation may indicate that catalase activity is low or not present at all in eggs. A complete developmental profile of catalase activities has not been reported.

The herbicide 3-amino-1,2,4-triazole (AT) irreversibly binds to catalase, resulting in loss of activity. SAMIS *et al.* (1972) have fed this inhibitor to adult males to study the kinetics of reappearance of catalase. Following a 12 hrs feeding of AT, catalase levels are reduced by 96%. Reappearance of catalase activity commences following a short lag, and normal levels are restored in about 1 week. Assuming that AT is cleared from the animal quickly and binding is irreversible in *Drosophila*, these kinetics can be used to estimate the rate of synthesis of catalase. It would be most interesting to compare the rates of catalase synthesis following AT treatment to the rates of synthesis which occur at the onset of catalase appearance during development.

Acetylcholinesterase (EC 3.1.1.7)

Acetylcholinesterase activity in *Drosophila* was initially studied by POULSON and BOELL (1946). It was observed that the activity in eggs is low, and increased during embryogenesis as the development of the nervous system proceeds. A correlation was observed between the volume of the nervous system and cholinesterase levels. The expected association between nervous tissue and cholinesterase was also observed by TUNNICLIFF *et al.* (1969), who found that the heads of adult flies contained high levels of cholinesterase. Recently, DEWHURST *et al.* (1970) have conducted a more extensive analysis of the levels of acetylcholinesterase found during development. When expressed as enzyme activity per individual, levels of acetylcholinesterase are seen to be fairly low in eggs with activity gradually increasing throughout larval life. An abrupt rise is seen in pupae several days before emergence. The maximum activity is obtained at the time of emergence, and remains more or less at this high level in adults. This profile differs significantly from that of a related enzyme choline acetyltransferase, (see below). DEWHURST *et al.* have also noted that 40% of the cholinesterase is found in the head.

MITCHELL *et al.* (1971) have observed that *Drosophila* acetylcholinesterase is quite sensitive to inhibition by the peptide, melittin, which is extracted from bee venom. This may explain partially the toxicity of this peptide to *Drosophila.*

Choline Acetyltransferase

The developmental history of choline acetyltransferase has also been studied by DEWHURST *et al.* (1970). Low activity is found in eggs, which then increases slightly during larval growth in a manner similar to acetylcholinesterase. However, its regulation during pupal life is quite different than the esterase. Transferase levels decline at pupation and remain low until just before emergence, when there is an abrupt rise in activity (70-fold). Maximal levels are found in young adults, and activity remains constant during adult life.

Carnitine Acetyltransferase (EC 2.3.1.7)

The association of carnitine acetyltransferase with normal sperm development has been studied by GEER and NEWBURGH (1970). While the function of this enzyme is not clear, it apparently is associated with the normal development of the sperm mitochondria. Dissection of testes has shown that the highest specific activity of carnitine acetyltransferase is found in the seminal vesicle, which is the area in which mature sperm are found. High enzyme levels are also found in isolated sperm. Lower activities are found in the regions of the testes containing cells at earlier stages of spermatogenesis.

Males which are sterile through being fed a diet substituting choline for carnitine, have low levels of carnitine acetyltransferase. Males which have the genetic constitution XO are also sterile, and also have reduced enzyme levels. Carnitine-fed and XO males both lack mature sperm in their testes. Examination of XO males which have different X-Y translocations indicate that the long arm of the Y-chromosome must be present for maximal carnitine acetyltransferase activity. The effect of induced or genetic sterility is specific for the sperm cells, since in these situations the levels of carnitine acetyltransferases found in the head and thorax is similar to controls.

GEER and NEWBURGH (1970) suggest that the low levels of carnitine acetyltransferase in the sterile situations are the result of failure of proper mitochondrial development during late spermatogenesis. Abnormal mitochondrial development is a common denominator in these sterile situations, which result in low enzyme levels. In addition, carnitine acetyltransferase is a known mitochondrial enzyme in other systems.

In a subsequent study, GEER *et al.* (1972) examined the metabolic aspects of energy generation during spermatogenesis. They measured the activity of the enzymes of the glycolytic cycle, citrate synthetase, NAD-IDH and MDH as représentative enzymes of the Krebs cycle, the NADP-dependent enzymes G6PDH, 6PGDH, NADP-IDH and malic enzyme, L-alanine aminotransferase and L-aspartate amino transferase as representatives of amino acid metabolizing enzymes and acetyl-Co-A carboxylase as a representative of fatty acid metabolism. The activities of these enzymes were compared in the testes of 1, 4, 6, and 10 day adult males. Comparisons were also made of these enzymes during thoracic maturation. Glycolytic enzymes, Krebs cycle enzymes and the amino transferases increased during this period, and appear to be important in late spermatogenesis. The NADP-dependent enzymes and acetyl CoA carboxylase seem to be relatively more important during the earlier stages of spermatogenesis.

References

ABE,K.: Genetical and biochemical studies on amylase in *Drosophila melanogaster*. Jap. J. Genet. **33**, 138–145 (1958).

ASHBURNER,M.: A prodromus to the genetic analysis of puffing in *Drosophila*. Cold Spring Harbor. Symp. Quant. Biol. **35**, 533–538 (1970a).

ASHBURNER,M.: Function and structure of polytene chromosomes during insect development. Advan. Insect Physiol. **7**, 1–95 (1970b).

AYALA,F.J., POWELL,J.R.: Enzyme variability in the *Drosophila willistoni* group. VI. Levels of polymorphism and the physiological function of enzymes. Biochem. Genet. **70**, 331–345 (1972).

AYALA,F.J., POWELL,J.R., TRACEY,M.L., MOURAO,C.A., PEREZSALAS,S.: Enzyme variability in the *Drosophila willistoni* group. IV. Genic variation in natural populations of *Drosophila willistoni*. Genetics **70**, 113–139 (1972).

BAGLIONI,C.: Genetic control of tryptophan peroxidase-oxidase in *Drosophila melanogaster*. Nature **184**, 1084–1085 (1959).

BAGLIONI,C.: The genetic control of tryptophan pyrrolase in *Drosophila melanogaster* and *Drosophila virilis*. Heredity **15**, 87–96 (1960).

BAHN,E.: Crossing over in the chromosomal region determining amylase isozymes in *Drosophila melanogaster*. Hereditas **58**, 1–12 (1968).

BAHN,E.: Cytogenetical localization of the amylase region in *Drosophila melanogaster* by means of translocations. Hereditas **67**, 75–78 (1971a).

BAHN,E.: Position-effect variegation for an isoamylase in *Drosophila melanogaster*. Hereditas **67**, 79–82 (1971b).

BAILLIE,D.L., CHOVNICK,A.: Studies on the genetic control of tryptophan pyrrolase in *Drosophila melanogaster*. Mol. Gen. Genet. **112**, 341–353 (1971).

BAKER,B.S.: The maternal and zygotic control of development by *cinnamon*, a new mutant in *Drosophila melanogaster*. Develop. Biol. **33**, 429–440 (1973).

BALLANTYNE,G.H., CHOVNICK,A.: Gene conversion in higher organisms: Non-reciprocal recombination events at the *rosy* cistron in *Drosophila melanogaster*. Genet. Res. (Camb.) **17**, 139–149 (1971).

BARTHELMESS,I.B., ROBERTSON,F.E.: The quantitative relations between variations in red eye pigment and related pteridine compounds in *Drosophila melanogaster*. Genet. Res. (Camb.) **15**, 65–86 (1970).

BEADLE,G.W.: Development of eye colors in *Drosophila:* Fat bodies and malpighian tubes in relation to diffusible substances. Genetics **22**, 587–611 (1937a).

BEADLE,G.W.: The development of eye color in *Drosophila* as studied by transplantation. Am. Naturalist **71**, 120–126 (1937b).

BEADLE,G.W.: Development of eye colors in *Drosophila:* fat bodies and malpighian tubes as sources of diffusible substances. Proc. Nat. Acad. Sci. U.S. **23**, 146–152 (1937c).

BEADLE,G.W., EPHRUSSI,B.: The differentiation of eye pigments in *Drosophila* as studied by transplantation. Genetics **21**, 225–247 (1936).

BEADLE,G.W., TATUM,E.L.: Experimental control of development and differentiation. Genetic control of developmental reactions. Am. Naturalist **75**, 107–116 (1941).

BEADLE,G.W., TATUM,E.L., CLANCY,C.W.: Food level in relation to rate of development and eye pigmentation in *Drosophila melanogaster*. Biol. Bull. **75**, 447–462 (1938).

BEADLE,G.W., TATUM,E.L., CLANCY,C.W.: Development of eye colors in *Drosophila:* Production of v^+ hormone by fat bodies. Biol. Bull. **77**, 407–414 (1939).

BECKMAN, L., JOHNSON, F. M.: Variations of larval alkaline phosphatase controlled by *Aph* alleles in *Drosophila melanogaster*. Genetics **49**, 829–835 (1964a).

BECKMAN, L., JOHNSON, F. M.: Esterase variations in *Drosophila melanogaster*. Hereditas **51**, 212–220 (1964b).

BECKMAN, L., JOHNSON, F. M.: Genetic control of aminopeptidases in *Drosophila melanogaster*. Hereditas **51**, 221–230 (1964c).

BECKMAN, L., JOHNSON, F. M.: Genetic variations of phosphatases in larvae of *Drosophila melanogaster*. Nature **201**, 32 (1964d).

BELL, J. B., MACINTYRE, R. J., OLIVIERI, A. P.: Induction of null-activity mutants for the acid phosphatase-1 gene in *Drosophila melanogaster*. Biochem. Genet. **6**, 205–216 (1972).

BENZER, S.: XIII International Congress of Genetics. Genetics. In press (1974).

BERGER, E. M.: A comparison of gene-enzyme variation between *Drosophila melanogaster* and *D. simulans*. Genetics **66**, 677–683 (1970).

BILLETT. F., COUNCE, S. J.: The β-glucuronidase content of *Drosophila* embryos. Exp. Cell. Res. **13**, 427–428 (1957).

BODENSTEIN, D., SACKTOR, B.: Cytochrome *c* oxidase activity during the metamorphosis of *Drosophila virilis*. Science **116**, 299–300 (1952).

BOND, P. A., SANG, J. H.: Glutamate dehydrogenase of *Drosophila* larvae. J. Insect. Physiol. **14**, 341–359 (1968).

BONI, P., DELERMA, B., PARISI, G.: Effects of the Inhibitor of Xanthine Dehydrogenase, 4-hydroxypyrazolo (3,4d) pyrimidine (or HPP) on the red eye pigments of *Drosophila melanogaster*. Experientia **23**, 186–187 (1967).

BONSE, A.: Über das Auftreten von Pterinen, Tryptophan und dessen Derivate in verschiedenen Organen der *Mutanta white* von *Drosophila melanogaster*. Z. Naturforsch. **24B**, 128–131 (1969).

BORACK, L. I., SOFER, W.: *Drosophila* β-L-hydroxy acid dehydrogenase, purification and properties. J. Biol. Chem. **246**, 5345–5350 (1971).

BOYD, J. B.: *Drosophila* deoxyribonucleases. I. Variation of deoxyribonucleases in *Drosophila melanogaster*. Biochim. Biophys. Acta **171**, 103–112 (1969).

BOYD, J. B.: Characterization of an alkaline active deoxyribonuclease from the prepupal salivary gland of *Drosophila hydei*. Biochim. Biophys. Acta **209**, 339–348 (1970a).

BOYD, J. B.: Characterization of an acid active deoxyribonuclease from the larval salivary gland of *Drosophila hydei*. Biochim. Biophys. Acta **209**, 349–356 (1970b).

BOYD, J. B., BERENDES, H. D., BOYD, H.: Mass preparation of nuclei from the larval salivary glands of *Drosophila hydei*. J. Cell. Biol. **38**, 369–376.

BOYD, J. B., BOYD, H.: Deoxyribonucleases of the organs of *Drosophila hydei* at the onset of metamorphosis. Biochem. Genet. **4**, 447–459 (1970).

BOYD, J. B., MITCHELL, H. K.: Identification of deoxyribonucleases in polyacrylamide gel following their separation by disc electrophoresis. Anal. Biochem. **13**, 28–42 (1965).

BRENNER-HOLZACH, O., LEUTHARDT, F.: Über die Aldolase aus *Drosophila melanogaster*. Helv. Chim. Acta **50**, 1366–1372 (1967).

BRENNER-HOLZACH, O., LEUTHARDT, F.: Reinigung und Kristallisation der Aldolase aus *Drosophila melanogaster*. Helv. Chim. Acta **51**, 1130–1133 (1968).

BRENNER-HOLZACH, O., LEUTHARDT, F.: Physikalisch-chemische, chemische und immunologische Eigenschaften der FDP-Aldolase aus *Drosophila melanogaster*. Helv. Chim. Acta **52**, 1273–1281 (1969).

BRITTEN, R. J., DAVIDSON, E. H.: Gene regulation for higher cells: a theory. Science **165**, 349–357 (1969).

BRUNET, P. C. J.: Tyrosine metabolism in insects. Ann. N.Y. Acad. Sci. **100**, 1020–1034 (1963).

BRUNET, P. C. J.: Sclerotins. Endeavour **26**, 68–74 (1967).

BUTENANDT, A., BIEKERT, E., LINZEN, B.: Über Ommochrome. VII. Mitteilung: Modellversuche zur Bildung des Xanthommatins *in vivo*. Hoppe Seylers Z. Physiol. Chem. **305**, 284–289 (1956).

BUTENANDT, A., SCHAFER, W.: Ommochromes. In: GORE, T. S., JOSHI, B. S., SUNTHANKAR, S. V., TILAK, B. D. (Eds.): Recent progress in the chemistry of natural and synthetic colouring matters, pp. 13–33. New York: Academic Press 1962.

CHAUHAN, N. S., ROBERTSON, F. W.: Quantitative inheritance of red eye pigment in *Drosophila melanogaster*. Genet. Res. (Camb.) **8**, 143–164 (1966).

CHEN, P. S.: Biochemical aspects of insect development. Basel-München-Paris-London-New York-Sydney: Karger 1971.

CHEN, P. S., HADORN, E.: Zur Stoffwechselphysiologie der Mutante *letal-meandor (lme)* von *Drosophila melanogaster*. Rev. Suisse Zool. **62**, 338–347 (1955).

CHOVNICK, A.: Genetic organization in higher organisms. Proc. Roy. Soc. (London), Ser. **B 164**, 198 (1966).

CHOVNICK, A., BALLANTYNE, G. H., BAILLIE, D. L., HOLM, D. G.: Gene conversion in higher organisms: half tetrad analysis of recombination within the *rosy* cistron of *Drosophila melanogaster*. Genetics **66**, 315–329 (1970).

CHOVNICK, A., BALLANTYNE, G. H., HOLM, D. A.: Studies on gene conversion and its relationship to linked exchange in *Drosophila melanogaster*. Genetics **69**, 179–209 (1971).

CHOVNICK, A., FINNERTY, V., SCHALET, A., DUCK, P.: Studies on the genetic organization of higher organisms: L. analysis of a complex gene in *Drosophila melanogaster*. Genetics **62**, 145–160 (1969).

CHOVNICK, A., SCHALET, A., KERNAGHAN, R. P., KRAUSS, M.: The *rosy* cistron in *Drosophila melanogaster*: genetic fine structure analysis. Genetics **50**, 1245–1259 (1964).

CHOVNICK, A., SCHALET, A., KERNAGHAN, R. P., TALSMA, J.: The resolving power of genetic fine structure analysis in higher organisms as exemplified by *Drosophila*. Am Naturalist **96**, 281–296 (1962).

CHRISTOPHER, C. W., JONES, M. E., STAFFORD, D. W.: Phenylalanine tRNA synthetase from *Drosophila melanogaster*. I. L-phenylalanine activation optima for pH, temperature and Mg^{2+} estimation of molecular weight. Biochim. Biophys. Acta **228**, 682–687 (1971).

COHEN, L. H., GOTCHEL, B. V.: Histones of polytene and nonpolytene nuclei of *Drosophila melanogaster*. J. Biol. Chem. **246**, 1841–1848 (1971).

COLLINS, J. F., DUKE, E. J., GLASSMAN, E.: Nutritional control of xanthine dehydrogenase. I. The effect in adult *Drosophila melanogaster* of feeding a high protein diet to larvae. Biochim. Biophys. Acta **208**, 294–303 (1970).

COLLINS, J. F., DUKE, E. J., GLASSMAN, E.: Multiple molecular forms of xanthine dehydrogenase and related enzymes. IV. The relationship of aldehyde oxidase to xanthine dehydrogenase. Biochem. Gen. **5**, 1–13 (1971).

COLLINS, J. F., GLASSMAN, E.: A third locus *(lpo)* affecting pyridoxal oxidase in *Drosophila melanogaster*. Genetics **62**, 833–839 (1969).

COÚNCE, S. J., WADDINGTON, C. H. (Eds.): Developmental systems-insects. New York: Academic Press 1972–1973.

COURTRIGHT, J. B.: Polygenic control of aldehyde oxidase in *Drosophila*. Genetics **57**, 25–39 (1967).

COURTRIGHT, J. B., IMBERSKI, R. B., URSPRUNG, H.: The genetic control of alcohol dehydrogenase and octanol dehydrogenase isozymes in *Drosophila*. Genetics **54**, 1251–1260 (1966).

DEJONG, G., HOORN, A. J. W., THORIG, G. E. W., SCHARLOO, W.: Frequencies of amylase variants in *Drosophila melanogaster*. Nature **238**, 453–454 (1972).

DEMEREC, M.: The biology of *Drosophila*. New York: Wiley 1950.

DEWHURST, S. A., MCCAMAN, R. E., KAPLAN, W. D.: The time course of development of acetylcholinesterase and choline acetyltransferase in *Drosophila melanogaster*. Biochem. Genet. **4**, 499–508 (1970).

DICKINSON, W. J.: Genetics and developmental regulation of aldehyde oxidase in *Drosophila melanogaster*. Genetics **60**, 173 (1968).

DICKINSON, W. J.: Developmental genetics of aldehyde oxidase in *Drosophila melanogaster*. Thesis, The Johns Hopkins University 1969.

DICKINSON, W. J.: The genetics of aldehyde oxidase in *Drosophila melanogaster*. Genetics **66**, 487–496 (1970).

DICKINSON, W. J.: Aldehyde oxidase in *Drosophila melanogaster*: A system for genetic studies on developmental regulation. Develop. Biol. **26**, 77–86 (1971).

DICKINSON, W. J.: A genetic locus affecting the developmental expression of an enzyme in *Drosophila*. Developmental Biology. In press (1975).

DOANE, W. W.: Disc electrophoresis of α-amylase isozymes in *Drosophila melanogaster*. Am. Zool. **5**, 697 (1965).

DOANE, W. W.: Quantitation of amylases in *Drosophila* separated by acrylamide gel electrophoresis. J. Exp. Zool. **164**, 363–378 (1967a).
DOANE, W. W.: Cytogenetic and biochemical studies of amylases in *Drosophila melanogaster*. Am. Zool. 7, 780 (1967b).
DOANE, W. W.: *Drosophila*. In: WILT, F., WESSELS, N. (Eds.): Methods in developmental biology, pp. 219–244. New York: T. Y. Crowell 1967c.
DOANE, W. W.: Amylase variants in *Drosophila melanogaster*: Linkage studies and characterization of enzyme extracts. J. Exp. Zool. **171**, 321–342 (1969a).
DOANE, W. W.: *Drosophila* amylases and problems in cellular differentiation. In: HANLEY, E. W. (Ed.): Problems in Biology: RNA in Development. Salt Lake City: University of Utah Press 1969b.
DOANE, W. W.: X-ray induced deficiencies of the *Amylase* locus in *Drosophila hydei*. Isozyme Bull. **4**, 46–48 (1971).
DOANE, W. W., ABRAHAM, I., KOLAR, M. M., MARTENSON, R. E., DEIBLER, G. E.: Purified *Drosophila* α-amylase isozymes: Genetical, biochemical and molecular characterization. In: Third International Isozyme Conference (MARKERT, C. L., ed.). In Press (1974).
DOANE, W. W., KOLAR, M. M.: Molecular weight of the polypeptide chain of α-amylase from an *Amy*^{7} strain of *Drosophila hydei*. Isozyme Bull. 7 (1973).
DOANE, W. W., KOLAR, M. M., SMITH, P. M.: Purification of α-amylase from *Drosophila*. Genetics **74**, 565 (1972).
DUNN, G. R., WILSON, T. G., JACOBSON, K. B.: Age-dependent changes in alcohol dehydrogenase in *Drosophila*. J. Exp. Zool. **171**, 185–190 (1969).
EPHRUSSI, B.: Analysis of eye color differentiation in *Drosophila*. Cold Spring Harbor Symp. Quant. Biol. **10**, 40–48 (1942).
EPHRUSSI, B., BEADLE, G. W.: A technique of transplantation for *Drosophila*. Am. Naturalist **70**, 218–225 (1936).
ESPOSITO, V. M., ULRICH, V.: Patterns of acid phosphatase in developing *Drosophila*. Genetics **54**, 334 (1966).
FABER, J., HADORN, E.: Non-autonomous pteridine formation induced by implantation of wild-type Malpighian tubules in two different lines of the mutant *maroon-like (ma-l)* of *Drosophila melanogaster*. Z. Vererbungslehre **94**, 242–248 (1963).
FALK, D. R., NASH, D.: Nutritionally supplementable mutations in *Drosophila melanogaster*. Genetics **74**, 576 (1973).
FARNSWORTH, M. W.: Growth and cytochrome c oxidase activity in larval stages of the *Minute (2) 1*^{2} mutant of *Drosophila*. J. Exp. Zool. **157**, 345–352 (1964).
FARNSWORTH, M. W.: Growth and cytochrome c oxidase activity in *Minute* mutants of *Drosophila*. J. Exp. Zool. **160**, 355–362 (1965a).
FARNSWORTH, M. W.: Oxidative phosphorylation in the *Minute* mutants of *Drosophila*. J. Exp. Zool. **160**, 363–368 (1965b).
FINNERTY, V., CHOVNICK, A.: Studies on genetic organization in higher organisms III. Confirmation of the single cistron-allele complementation model of organization of the *maroon-like* region of *Drosophila melanogaster*. Genet. Res. (Camb.) **15**, 51–355 (1970).
FINNERTY, V., DUCK, P., CHOVNICK, A.: Studies on genetic organization in higher organisms. II. Complementation and fine structure of the *maroon-like* locus of *Drosophila melanogaster*. Proc. Nat. Acad. Sci. U.S. **65**, 939–940 (1970).
FORREST, H.: The ommochromes. In: GORDON, M. (Ed.): Pigment cell biology, pp. 619–628. New York: Academic Press 1959.
FORREST, H., Pteridines: structure and metabolism. In: FLORKIN, M., MASON, H. S. (Eds.): Comparative biochemistry, Vol. IV, pp. 615–641. New York: Academic Press 1962.
FORREST, H., GLASSMAN, E., MITCHELL, H. K.: Conversion of 2-amino-4-hydroxy-pteridine to isoxanthopterin in *D. melanogaster*. Science **124**, 725–726 (1956).
FORREST, H., HANLY, E. W., LAGOWSKI, J. M.: Biochemical differences between the mutants *rosy-2* and *maroon-like* of *Drosophila melanogaster*. Genetics **40**, 1455–1463 (1961).
FORREST, H., HATFIELD, D., VAN BAALEN, C.: Characterization of a second yellow compound from *Drosophila melanogaster*. Nature **183**, 1269–1270 (1959).
FORREST, H., MITCHELL, H. K.: Pteridines from *Drosophila*. I. Isolation of a yellow pigment. J. Am. Chem. Soc. 76, 5656–5658 (1954a).

FORREST, H., MITCHELL, H. K.: Pteridines from *Drosophila*. II. Structure of the yellow pigment. J. Am. Chem. Soc. **76**, 5658–5662 (1954b).

FOX, D. J.: The soluble citric acid cycle enzymes of *Drosophila melanogaster*. I. Genetics and ontogeny of NADP-linked isocitrate dehydrogenase. Biochem. Genet. **5**, 69–80 (1971).

FOX, D. J., CONSCIENCE-EGLI, M., ABACHERLI, E.: The soluble citric acid cycle enzymes of *Drosophila melanogaster*. II. Tissue and intracellular distribution of aconitase and NADP-dependent isocitrate dehydrogenase. Biochem. Genet. **7**, 163–175 (1972).

FRISTROM, J. W.: The developmental biology of *Drosophila*. Ann. Rev. Genet. **4**, 325–346 (1970).

FRISTROM, J. W., MITCHELL, H. K.: The preparative isolation of imaginal discs from larvae of *Drosophila melanogaster*. J. Cell. Biol. **27**, 445–448 (1965).

GARCIA-BELLIDO, A., MERRIAM, J. R.: Cell lineage of the imaginal discs in *Drosophila* gynandromorphs. J. Exp. Zool. **170**, 61–76 (1969).

GEER, B. W., MARTENSEN, D. V., DOWNING, B. C., MUZYKA, G. S.: Metabolic changes during spermatogenesis and thoracic tissue maturation in *Drosophila hydei*. Develop. Biol. **28**, 390–406 (1972).

GEER, B. W., NERBURGH, R. W.: Carnitine acetyltransferase and spermatozoan development in *Drosophila melanogaster*. J. Biol. Chem. **245**, 71–79 (1970).

GEIGER, H. R., MITCHELL, H. K.: Salivary gland function in phenol oxidase production in *Drosophila melanogaster*. J. Insect. Physiol. **12**, 747–754 (1966).

GEORGIEV, G. P.: The structure of transcriptional units in eukaryotic cells. In: MOSCONA, A. A., MONROY, A. (Eds.): Current topics in developmental biology, Vol. 7, pp. 1–60. New York: Academic Press 1972.

GHOSH, D., FORREST, H. S.: Inhibition of tryptophan pyrrolase by some naturally occurring pteridines. Arch. Biochem. Biophys. **120**, 578–582 (1967a).

GHOSH, D., FORREST, H. S.: Enzymatic studies on the hydroxylation of kynurenine in *Drosophila melanogaster*. Genetics **55**, 423–431 (1967b).

GILLESPIE, J. H., KOJIMA, K.: The degree of polymorphism in enzymes involved in energy production compared to that in nonspecific enzymes in two *Drosophila ananassae* populations. Proc. Nat. Acad. Sci. U.S. **61**, 582–585 (1968).

GLASSMAN, E.: Kynurenine formamidase in mutants of *Drosophila*. Genetics **41**, 566–574 (1956).

GLASSMAN, E.: Convenient assay of xanthine dehydrogenase in a single *Drosophila melanogaster*. Science **137**, 990–991 (1962a).

GLASSMAN, E.: Unexpected presence of xanthine dehydrogenase in combined extracts of *maroon-like* and *rosy* eye color mutants of *Drosophila melanogaster: a* case of *in vitro* complementation between non-allelic genes. Genetics **47**, 954 (1962b).

GLASSMAN, E.: *In vitro* complementation between non-allelic *Drosophila* mutants deficient in xanthine dehydrogenase. Proc. Nat. Acad. Sci. U.S. **48**, 1491–1497 (1962c).

GLASSMAN, E.: Genetic regulation of xanthine dehydrogenase in *Drosophila melanogaster*. Federation Proc. **24**, 1243–1251 (1965).

GLASSMAN, E.: Complementation *in vitro* between non-allelic, *Drosophila* mutants deficient in xanthine dehydrogenase. III. Observations on heat stabilities. Biochim. Biophys. Acta **117**, 342–350 (1966).

GLASSMAN, E., KARAM, J. D., KELLER, E. C.: Differential response to gene dosage experiments involving the two loci which control xanthine dehydrogenase of *Drosophila melanogaster*. Z. Vererbungslehre **93**, 399–403 (1962).

GLASSMAN, E., KELLER, Jr., E. C., KARAM, J. D., MCLEAN, J., CATES, M.: *In vitro* complementation between non-allelic mutants deficient in xanthine dehydrogenase. II. The absence of the *ma-l*$^{+}$ factor in *lxd* mutant flies. Biochem. Biophys. Res. Commun. **17**, 242–247 (1964).

GLASSMAN, E., MCLEAN, J.: Maternal effect of *ma-l* on xanthine dehydrogenase of *Drosophila melanogaster*. II. Xanthine dehydrogenase activity during development. Proc. Nat. Acad. Sci. U.S. **48**, 1712–1718 (1962).

GLASSMAN, E., MITCHELL, H. K.: Mutants of *Drosophila melanogaster* deficient in xanthine dehydrogenase. Genetics **44**, 153–162 (1959).

GLASSMAN, E., PINKERTON, W.: Complementation at the *maroon-like* eyecolor locus of *Drosophila melanogaster*. Science **131**, 1810–1811 (1960).

GLASSMAN, E., SHINODA, T., DUKE, E. J., COLLINS, J. F.: Multiple molecular forms on xanthine dehydrogenase and related enzymes. Ann. N. Y. Acad. Sci. **151**, 263–273 (1968).

GLASSMAN, E., SHINODA, I., MOON, H. M., KARAM, J. D.: *In vitro* complementation between non-allelic *Drosophila* mutants deficient in xanthine dehydrogenase. IV. Molecular weights. J. Mol. Biol. **20**, 419–422 (1966).

GRAUBARD, M. A.: Tyrosinase in mutants of *Drosophila melanogaster*. J. Genet. **27**, 199–218 (1933).

GREEN, M. M.: Mutant isoalleles at the *vermilion* locus in *Drosophila melanogaster*. Proc. Nat. Acad. Sci. U.S. **38**, 300–305 (1952).

GREEN, M. M.: Pseudo-allelism at the *vermilion* locus in *Drosophila melanogaster*. Proc. Nat. Acad. Sci. U.S. **40**, 92–99 (1954).

GREGG, T. G., SMUCKER, L. A.: Pteridines and gene homologies in the eye color mutants of *Drosophila hydei* and *Drosophila melanogaster*. Genetics **52**, 1023–1034 (1965).

GRELL, E. H.: The dose effect of $ma\text{-}l^+$ and ry^+ on xanthine dehydrogenase activity in *Drosophila melanogaster*. Genetics **47**, 950 (1962a).

GRELL, E. H.: The dose effect of $ma\text{-}l^+$ and ry^+ on xanthine dehydrogenase activity in *Drosophila melanogaster*. Z. Vererbungslehre **93**, 371–377 (1962b).

GRELL, E. H.: Electrophoretic variants of a-glycerophosphate dehydrogenase in *Drosophila melanogaster*. Science **158**, 1319–1320 (1967).

GRELL, E. H., JACOBSON, K. B., MURPHY, J. B.: Alcohol dehydrogenase in *Drosophila melanogaster:* Isozymes and genetic variants. Science **149**, 80–82 (1965).

GRELL, E. H., JACOBSEN, K. B., MURPHY, J. B.: Alterations of genetic material for analysis of alcohol dehydrogenase isozymes of *Drosophila melanogaster*. Ann. N. Y. Acad. Sci. **151**, 441–455 (1968).

HADORN, E.: Patterns of biochemical and developmental pleiotropy. Cold Spring Harbor Symp. Quant. Biol. **21**, 363–373 (1956).

HADORN, E.: Contribution to the physiological and biochemical genetics of pteridines and pigments in insects. Int. Congr. Genet., 19th Congr. **1**, 337–354 (1958).

HADORN, E., GRAF, G. E.: Weitere Untersuchungen über den nicht-autonemen Pterinstoffwechsel der Mutante *rosy* von *Drosophila melanogaster*. Zool. Anzeiger **160**, 231–243 (1958).

HADORN, E., MITCHELL, H. K.: Properties of mutants of *Drosophila melanogaster* and changes during development as revealed by paper chromatography. Proc. Nat. Acad. Sci. U.S. **37**, 650–665 (1951).

HADORN, E., SCHWINCK, I.: A mutant of *Drosophila* without iso-xanthopterine which is non-autonomous for the red eye pigments. Nature **177**, 940-941 (1956a).

HADORN, E., SCHWINCK, I.: Fehlen von Isoxanthopterin und Nicht-Autonomie in der Bildung der roten Augenpigmente bei einer Mutante *(rosy²)* von *Drosophila melanogaster*. Z. Vererbungslehre **87**, 528–553 (1956b).

HADORN, E., ZIEGLER-GUNDER, I.: Untersuchungen zur Entwicklung, Geschlechtsspezifität, und Phänogenetischen Autonomie der Augen-Pterine verschiedener Genotypen von *Drosophila melanogaster*. Z. Vererbungslehre **89**, 221–234 (1958).

HANDSCHIN, G.: Entwicklungs- und organspezifisches Verteilungsmuster der Pterine bei einem Wildstamm und bei der Mutante $rosy^2$ von *Drosophila melanogaster*. Develop. Biol. **3**, 115–149 (1961).

HARNLY, M. H., EPHRUSSI, B.: Development of eye colors in *Drosophila:* time of action of body fluid on cinnabar. Genetics **22**, 393–401 (1937).

HARPER, R. A., ARMSTRONG, F. B.: Alkaline phosphatase of *Drosophila melanogaster*. I. Partial purification and characterization. Biochem. Genet. **6**, 75–82 (1972).

HENDERSON, A. S., GLASSMAN, E.: A possible storage form for tyrosinase substrates in *Drosophila melanogaster*. J. Insect. Physiol. **15**, 2345–2355 (1969).

HERDING, S.: Aldehyde oxidase in *Drosophila* adults and larvae. Thesis, Reed College 1970.

HJORTH, J. P.: A phosphoglucomutase locus in*Drosophila melanogaster*. Hereditas **64**, 146–148 (1970).

HODGE, L. D., GLASSMAN, E.: Purine catabolism in *Drosophila melanogaster*. I. Reactions leading to xanthine dehydrogenase. Biochim. Biophys. Acta **149**, 335–343 (1967).

HODGE, L. D., GLASSMAN, E.: Purine catabolism in *Drosophila melanogaster*. II. Guanine deaminase, inosine phosphorylase and adenosine deaminase activities in mutants with altered xanthine dehydrogenase activities. Genetics **57**, 571–577 (1967).

HODGETTS, R. B., KONOPKA, R. J.: Tyrosine and catecholamine metabolism in wild type *Drosophila melanogaster* and a mutant, *ebony*. J. Insect. Physiol. **19**, 1211–1220 (1973).

HORIKAWA, M., LING, L. L., FOX, A. S.: Effects of substrates on gene-controlled enzyme activities in cultured embryonic cells of *Drosophila*. Genetics **55**, 569–583 147, autocatalytic (1967).

HOROWITZ, N. H., FLING, M.: The autocatalytic production of tyrosinase in extracts of *Drosophila melanogaster*. In: MCELROY, W. D., GLASS, B. (Eds.): Amino acid metabolism. Baltimore: Johns Hopkins Univ. Press 1955.

HOTTA, Y., BENZER, S.: Mapping of behavior in *Drosophila* mosaics. Nature **240**, 527–535 (1972).

HUANG, S. L., SINGH, M., KOJIMA, K.: A study of frequency-dependent selection observed in the *esterase-6* locus of *Drosophila melanogaster* using a conditioned media method. Genetics **68**, 97–104 (1971).

HUBBY, J. L.: A mutant affecting pteridine metabolism in *Drosophila melanogaster*. Genetics **47**, 109–114 (1962).

HUBBY, J. L., LEWONTIN, R. C.: A molecular approach to the study of genic heterozygosity in natural populations. I. The number of alleles at different loci in *Drosophila pseudoobscura*. Genetics **54**, 577–594 (1966).

HUBBY, J. L., NARISE, S.: Protein differences in *Drosophila*. III. Allelic differences and species differences in *in vitro* hybrid enzyme formation. Genetics **57**, 291–300 (1967).

HUBBY, J. L., THROCKMORTON, L. H.: Evolution and pteridine metabolism in the genus *Drosophila*. Proc. Nat. Acad. U.S. **46**, 65–78 (1960).

HUBER, R. E., LEFEBVRE, Y. A.: The purification and some properties of souble trehalase and sucrase from *Drosophila melanogaster*. Can. J. Biochem. Physiol. **49**, 1155–1164 (1971).

HUNTER, L. R., MARKERT, C. L.: Histochemical demostration of enzymes separated by zone electrophoresis in starch gels. Science **125**, 1294–1295 (1957).

IMBERSKI, R. B.: Isozymes of aldehyde oxidase in *Drosophila hydei*. Genetics **63**, 530 (1971).

IMBERSKI, R. B., SOFER, W. H., URSPRUNG, H.: *Drosophila* alcohol dehydrogenase isozymes: Identity of molecular size. Experientia **24**, 504–505 (1968).

JACOBSON, K. B.: Alcohol dehydrogenase of *Drosophila*:Interconversion of isozymes. Science **159**, 324–325 (1968).

JACOBSON, K. B.: Role of an isoacceptor transfer ribonucleic acid as an enzyme inhibitor: Effect on tryptophan pyrrolase of *Drosophila*. Nature New Biol. **231**, 17–19 (1971).

JACOBSON, K. B., MURPHY, J. B., HARTMAN, F. C.: Isoenzymes of *Drosophila* alcohol dehydrogenase. I. Isolation and interconversion of different forms. J. Biol. Chem. **245**, 1075–1083 (1970).

JACOBSON, K. B., MURPHY, J. B., KNOPP, J. A., ORTIZ, J. R.: Multiple forms of *Drosophila* alcohol dehydrogenase. III. Conversion of one form to another by nicotinamide adenine dinucleotide or acetone. Arch. Biochem. Biophys. **149**, 22–35 (1972).

JACOBSON, K. B., PFUDERER, P.: Interconversion of isoenzymes of *Drosophila* alcohol dehydrogenase. II. Physical characterization of the enzyme and its subunits. J. Biol. Chem. **245**, 3938–3944 (1970).

JANNING, W.: Aldehyde oxidase as a cell marker for internal organs in *Drosophila melanogaster*. Naturwissenschaften **59**, 516–517 (1972).

JANNING, W.: Entwicklungsgenetische Untersuchungen an Gynandern von *Drosophila melanogaster*. I. Die inneren Organe der Imago. Roux' Archiv **174**, 313–332 (1974a).

JANNING, W.: Entwicklungsgenetische Untersuchungen an Gynandern von *Drosophila melanogaster*. II. Der morphogenetische Anlageplan. Roux' Archiv **174**, 349–359 (1974b).

JELNES, J. E.: Identification of hexokinases and localization of a fructokinase and a tetrazolium oxidase locus in *Drosophila melanogaster*. Hereditas **67**, 291–293 (1971).

JOHNSON, F. M.: Sex-limited inheritance of some esterase variations in *Drosophila melanogaster*. Genetics **50**, 259 (1964a).

JOHNSON, F. M.: A recessive esterase deficiency in *Drosophila*. J. Heredity **55**, 76–78 (1964b).

JOHNSON, F. M.: Developmental differences of alkaline phosphatase zymograms from *Drosophila melanogaster* and *D. ananassae*. Nature **212**, 843–844 (1966a).

JOHNSON, F. M.: *Drosophila melanogaster:* Inheritance of a deficiency of alkaline phosphatase in larvae. Science **152**, 361–362 (1966b).

JOHNSON, F. M.: Isozyme polymorphisms in *Drosophila ananassae:* genetic diversity among island populations in the South Pacific. Genetics **68**, 77–95 (1971).

JOHNSON, F. M., BEALLE, S.: Isozyme variability in species of the genus *Drosophila*. V. Ejaculatory bulb esterases in *Drosophila* phylogeny. Biochem. Genet. **2**, 1–18 (1968).

JOHNSON, F. M., DENNISTON, C.: Genetic variation of alcohol dehydrogenase in *Drosophila melanogaster*. Nature **204**, 906–907 (1964).

JOHNSON, F. M., KANAPI, C. G., RICHARDSON, R. H., SAKAI, R. K.: Isozyme variability in species of the genus *Drosophila*. I. A multiple allelic isozyme system in *Drosophila busckii*: Inheritance and general considerations. Biochem. Genet. **1**, 35–40 (1967).

JOHNSON, F. M., KANAPI, C. G., RICHARDSON, R. H., WHEELER, M. R., STONE, W. S.: An operational classification of *Drosophila* esterases for species comparisons. Studies in Genetics. Morgan Centennial Issue, University of Texas Publication **6615**, 517–532 (1966a).

JOHNSON, F. M., KANAPI, C. G., RICHARDSON, R. H., WHEELER, M. R., STONE, W. S.: An analysis of polymorphisms among isozyme loci in dark and light *Drosophila ananassae* strains from American and Western Samoa. Proc. Nat. Acad. Sci. U.S. **56**, 119–125 (1966b).

JOHNSON, F. M., RICHARDSON, R. H., KAMBYSELLIS, M. P.: Isozyme variability in species of the Genus *Drosophila*. III. Qualitative comparison of the esterases of *D. aldrichi* and *D. mulleri*. Biochem. Genet. **1**, 234–249 (1968).

JOHNSON, F. M., SAKAI, R. K.: A leucine aminopeptidase polymorphism in *Drosophila buskii*. Nature **203**, 373–374 (1964).

JUDD, B. H., SHEN, M. W., KAUFMAN, T. C.: The anatomy and function of a segment of the X chromosome of *Drosophila melanogaster*. Genetics **71**, 139–156 (1972).

KAMBYSELLIS, M. P., JOHNSON, F. M., RICHARDSON, R. H.: Isozyme variability in species of the genus *Drosophila*. IV. Distribution of esterases in the body tissues of *D. aldrichi* and *D. mulleri* adults. Biochem. Genet. **1**, 249–265 (1968).

KARLSON, P., SEKERIS, C. E.: Biochemistry of insect metamorphosis. In: FLORKIN, M., MASON, H. S. (Eds.): Comparative biochemistry, Vol. VI, p. 221–236. New York: Academic Press 1964.

KAUFMAN, S.: Studies on tryptophan pyrrolase in *Drosophila melanogaster*. Genetics **47**, 807–817 (1962).

KAZAZIAN, H. H. JR.: Molecular size studies on 6-phosphogluconate dehydrogenase. Nature **212**, 197–198 (1966).

KAZAZIAN, H. H., JR., YOUNG, W. J., CHILDS, B.: X-linked 6-phosphogluconate dehydrogenase in *Drosophila:* Subunit associations. Science **150**, 1601–1602 (1965).

KELLER, E. C.: Qualitative differences in xanthine dehydrogenase activity in wild-type strains of *Drosophila melanogaster*. Z. Vererbungslehre **95**, 326–332 (1964).

KELLER, E. C., GLASSMAN, E.: A third locus *(lxd)* affecting xanthine dehydrogenase in *Drosophila melanogaster*. Genetics **49**, 663–668 (1964a).

KELLER, E. C., GLASSMAN, E.: Xanthine dehydrogenase differences in activity among *Drosophila* strains. Science **143**, 40–41 (1964b).

KELLER, E. C., GLASSMAN, E.: Selection for xanthine dehydrogenase in *Drosophila melanogaster*. J. Elisha Mitchell Sci. Soc. **80**, 130–133 (1964c).

KELLER, E. C., GLASSMAN, E.: Phenocopies of the *ma-l* and *ry* mutants of *Drosophila melanogaster*. Inhibition *in vivo* of xanthine dehydrogenase by 4-hydroxy pyrazola (3,4-d) pyrimidine. Nature **208**, 202–203 (1965).

KIKKAWA, H.: Further studies on the genetic control of amylase in *Drosophila melanogaster*. Jap. J. Genet. **33**, 382–387 (1960).

KIKKAWA, H.: An electrophoretic study on amylase in *Drosophila melanogaster*. Jap. J. Genet. **39**, 401–411 (1964).

KIKKAWA, H.: Biochemical genetics of proteolytic enzymes in *Drosophila melanogaster. 1. General considerations. Japan J. Genet.* **43**, 137–148 (1968).

KIKKAWA, H., ABE, K.: Genetic control of amylase in *Drosophila melanogaster*. Annot. Zool. Jap. **33**, 14–23 (1960).

KIMMEL, D. L., JR.: Tryptophan oxygenase and formylase activities during larval development of wild type *Drosophila melanogaster*. Am. Zoologist **9**, 603 (1969).

KIMMEL, D. L., JR.: Properties of kynurenine formamidase from adult wild type *Drosophila melanogaster*. Genetics **64**, supplement 33 (1970).

KING, J. C.: Differences in levels of xanthine dehydrogenase activity between inbred and outbred strains of *Drosophila melanogaster*. Proc. Nat. Acad. Sci. U.S. **64**, 891–896 (1969).

KNOPP, J. A., JACOBSON, K. B.: Multiple forms of *Drosophila* alcohol dehydrogenase. IV. Protein fluorescence studies. Arch. Biochem. Biphys. **149**, 36–41 (1972).

KNOWLES, B. B., FRISTROM, J. W.: The electrophoretic behaviour of ten enzyme systems in the larval integument of *Drosophila melanogaster*. J. Insect Physiol. **13**, 731–737 (1967).

KNUTSEN, C., SING, C. F., BREWER, G. J.: Hexokinase isozyme variability in *Drosophila robusta*. Biochem. Genet. **3**, 475–483 (1969).

KOJIMA, K., GILLESPIE, J., TOBARI, Y. M.: A profile of *Drosophila* species enzymes assayed by electrophoresis. I. Number of alleles, heterozygosities and linkage disequilibrium in glucose metabolizing systems and some other enzymes. Biochem. Genet. **4**, 627–637 (1970).

KOJIMA, K., SMOUSE, P., YANG, S., NAIR, P. S., BRNCIC, D.: Isozyme frequency patterns in *Drosophila pavani* associated with geographical and seasonable variables. Genetics **72**, 721–731 (1972).

KOJIMA, K., YARBROUGH, K. M.: Frequency—Dependent selection at the *esterase*-6 locus in *Drosophila melanogaster*. Proc. Nat. Acad. Sci. U.S. **57**, 645–649 (1967).

KOMMA, D. J.: Effect of sex transformation genes on glucose-6-phosphate dehydrogenase activity in *Drosophila melanogaster*. Genetics **54**, 497–503 (1966).

KOMMA, D. J.: Glucose 6-phosphate dehydrogenase in *Drosophila*: A sex-influenced electrophoretic variant. Biochem. Genet. **1**, 229–237 (1968 a).

KOMMA, D. J.: Glucose 6-phosphate dehydrogenase in *Drosophila*: sexual effects on structure. Biochem. Genet. **1**, 337–346 (1968 b).

KRUGELIS, E. J.: Distribution and properties of intracellular alkaline phosphatases. Biol. Bull. **90**, 220–223 (1946).

LAIRD, C. D.: DNA of *Drosophila* chromosomes. Ann. Rev. Genet. **7**, 177–204 (1973).

LEWIS, E. B.: Genetic control and regulation of developmental pathways. In: LOCKE, M. (Ed.): The role of chromosomes in development, pp. 231–252. New York: Academic Press 1964.

LEWIS, H. W., LEWIS, H. S.: Genetic control of dopa oxidase activity in *Drosophila melanogaster*. II. Regulating mechanisms and inter- and intra-strain heterogeneity. Proc. Nat. Acad. Sci. U.S. **47**, 78–86 (1961).

LEWIS, H. W., LEWIS, H. S.: Genetic regulations of dopa oxidase activity in *Drosophila*. Ann. N. Y. Acad. Sci. **100**, 827–839 (1963).

LEWONTIN, R. C.: Population Genetics. Ann. Rev. Genet. **7**, 1–17 (1973).

LEWONTIN, R. C., HUBBY, J. L.: A molecular approach to the study of genic heterozygosity in natural populations. II. Amount of variation and degree of heterozygosity in natural populations of *Drosophila pseudoobscura*. Genetics **54**, 595–609 (1966).

LINDSLEY, D. L., GRELL, E. H.: The genetic variations of *Drosophila melanogaster*. Carnegie Inst. (Wash.) Publ. **627** (1967).

LINZEN, B.: Zur Biochemie der Ommochrome. Naturwissenschaften **21 b**, 259–267 (1967).

LUCAS-LENARD, J., LIPMANN, F.: Protein synthesis. Ann. Rev. Biochem. **40**, 409–448 (1971).

LUCCHESI, J. C.: Dosage compensation in *Drosophila*. Ann. Rev. Genet. **7**, 225–237 (1973).

LUNAN, K. D., MITCHELL, H. K.: The metabolism of tyrosine-0-phosphate in *Drosophila*. Arch. Biochem. Biophys. **132**, 450–456 (1969).

MACINTYRE, R. J.: Locus of the structural gene for third larval instar alkaline phosphatase. DIS **41**, 62 (1966 a).

MACINTYRE, R. J.: The genetics of an acid phosphatase in *Drosophila melanogaster* and *D. simulans*. Genetics **53**, 461–474 (1966 b).

MACINTYRE, R. J.: A method for measuring activities of acid phosphatases separated by acrylamide gel electrophoresis. Biochem. Genet. **5**, 45–56 (1971 a).

MACINTYRE, R. J.: Evolution of acid phosphatase-1 in the genus *Drosophila* as estimated by subunit hybridization. 1: Methodology. Genetics **68**, 483–508 (1971 b).

MACINTYRE, R. J., DEAN, M. R.: Sub-units of acid phosphatase-1 in *Drosophila melanogaster*: Reversible dissociation *in vitro*. Nature **214**, 274–275 (1967).

MACINTYRE, R. J., WRIGHT, T. R.: Responses of esterase-6 alleles of *Drosophila melanogaster* and *D. simulans* to selection in experimental populations. Genetics **53**, 371–387 (1966).

MADHAVAN, K. M., CONSCIENCE-EGLI, F., SIEBER, F., URSPRUNG, H.: Farnesol metabolism in *Drosophila melanogaster:* Ontogeny and distribution of octanol dehydrogenase and aldehyde oxidase. J. Insect Physiol. **19**, 235–241 (1973).

MADHAVAN, K. M., FOX, D. J., URSPRUNG, H.: Developmental genetics of hexokinase isozymes in *Drosophila melanogaster*. J. Insect Physiol. **18**, 1523–1530 (1972).

MADHAVAN, K. M., URSPRUNG, H.: The genetic control of fumarate hydratase (fumarase) in *Drosophila melanogaster*. Mol. Gen. Genet. **120**, 379–380 (1973).

MAGDON, E.: Untersuchungen zur Thymidylatsynthetase bei *Drosophila melanogaster*. Z. Naturforsch. **24b**, 1069–1070 (1969).

MARKERT, C. L.: Lactate dehydrogenase isozymes: dissociation and recombination of subunits. Science **140**, 1329–1330 (1963).

MARZLUF, G. A.: Tryptophan pyrrolase of *Drosophila:* Partial purification and properties. Z. Vererbungslehre **97**, 10–17 (1965 a).

MARZLUF, G. A.: Enzymatic studies with the suppressor of *vermilion* of *Drosophila melanogaster*. Genetics **52**, 503–512 (1965 b).

MARZLUF, G. A.: Studies of trehalase and sucrase of *Drosophila melanogaster*. Arch. Biochem. Biophys. **134**, 8–18 (1969).

MCCAMAN, M. W., MCCAMAN, R. E., LEES, G. E.: Liquid cation exchange—a basis for sensitive radiometric assays for aromatic amino acid decarboxylases. Anal. Biochem. **45**, 242–252 (1972).

MCREYNOLDS, M. S.: Homogolous esterases in three species of the virilis group of *Drosophila*. Genetics **56**, 527–540 (1967).

MCREYNOLDS, M. S., KITTO, G. B.: Purification and properties of *Drosophila* malate dehydrogenases. Biochim. Biophys. Acta **198**, 165–175 (1970).

MEYER-TAPLICK, T., CHEN, P. S.: Zur Histologie normaler und letaler *(lme)* Larven von *Drosophila melanogaster*. Rev. Suisse Zool. **67**, 245–257 (1960).

MITCHELL, H. K.: Phenol oxidases and *Drosophila* development. J. Insect. Physiol. **12**, 755–765 (1966).

MITCHELL, H. K., GLASSMAN, E., HADORN, E.: Hypoxanthine in *rosy* and *maroon-like* mutants of *Drosophila melanogaster*. Science **129**, 268 (1959).

MITCHELL, H. K., LOWY, P. H., SARMIENTO, L., DICKSON, L.: Melittin: Toxicity to *Drosophila* and inhibition of acetylcholinesterase. Arch. Biochem. Biophys. **145**, 344–348 (1971).

MITCHELL, H. K., LUNAN, K. D.: Tyrosine-0-phosphate in *Drosophila*. Arch. Biochem. Biophys. **106**, 219–222 (1964).

MITCHELL, H. K., MITCHELL, A.: Mass culture and age selection in *Drosophila*. Drosophila Info. Ser. **39**, 135–137 (1964).

MITCHELL, H. K., WEBER, U. M.: *Drosophila* phenol oxidases. Science **148**, 964–965 (1965).

MITCHELL, H. K., WEBER, U. M., SCHAAR, G.: Phenol oxidase characteristics in mutants of *Drosophila melanogaster*. Genetics **57**, 357–368 (1967).

MITCHELL, H. K., WEBER-TRACY, U. M., SCHAAR, G.: Aspects of cuticle formation in *Drosophila melanogaster*. J. Exp. Zool. **176**, 429–444 (1971).

MIZIANTY, T. J., CASE, S. T.: Demonstration of genetic control of esterase—A in *Drosophila melanogaster*. J. Heredity **62**, 345–348 (1971).

MORGAN, T. H.: Sex limited inheritance in *Drosophila*. Science **32**, 120–122 (1910).

MORITA, T.: Purine catabolism in *Drosophila melanogaster*. Science **128**, 1135 (1958).

MORRISON, W. W., FRAJOLA, W. J.: The stimulation of tryptophan pyrrolase activity by RNA fractions isolated at various temperatures. Biochem. Biophys. Res. Commun. **17**, 597–602 (1964).

MUHAMMED, A., GONCALVES, J. M., TROSKO, J. E.: Deoxyribonuclease and deoxyribonucleic acid polymerase activity during *Drosophila* development. Develop. Biol. **15**, 23–32 (1967).

MUNZ, P.: Untersuchungen über die Aktivität der Xanthine Dehydrogenase in Organen und während der Ontogenese von *Drosophila melanogaster*. Z. Vererbungslehre **95**, 195–210 (1964).

MURRAY, R. F., JR., BALL, J. A.: Testis-specific and sex associated hexokinases in *Drosophila melanogaster*. Science **156**, 81–82 (1967).

NARISE, S., HUBBY, J. L.: Purification of esterase—5 from *Drosophila pseudoobscura*. Biochim. Biophys. Acta **122**, 281–288 (1966).

NASH, D., FALK, D. R.: Pyrimidine-requiring mutants of *Drosophila melanogaster*. Genetics **74**, 6191 (1973).

NORBY, S.: A specific nutritional requirement for pyrimidines in rudimentary mutants of *Drosophila melanogaster*. Hereditas **66**, 205–214 (1970).

NOVIKOFF, A. B.: The validity of histochemical phosphatase methods on the intracellular level. Science **113**, 320–325 (1951).

NOVITSKI, E., ASHBURNER, M. (Eds.): Biology of *Drosophila*. New York: Academic Press (in press).

O'BRIEN, S. J., MACINTYRE, R. J.: An analysis of gene-enzyme variability in natural populations of *Drosophila melanogaster* and *D. simulans*. Am. Naturalist **103**, 97–113 (1969).

O'BRIEN, S. J., MACINTYRE, R. J.: The α-glycerophosphate cycle in *Drosophila melanogaster*. I. Biochemical and developmental aspects. Biochem. Genet. **7**, 141–161 (1972a).

O'BRIEN, S. J., MACINTYRE, R. J.: The α-glycerophosphate in *Drosophila melanogaster* II. Genetic aspects. Genetics **71**, 127–138 (1972b).

OGONJI, G. O.: Bearing of genetic data on the interpretation of the subunit structure of octanol dehydrogenase of *Drosophila*. J. Exp. Zool. **178**, 513–522 (1971).

OHNISHI, E.: Activation of tyrosinase in *Drosophila virilis*. Annot. Zool. Jap. **27**, 188–193 (1954).

PAPPAS, P. W., RODRICK, G. E.: An electrophoretic study of lactate dehydrogenase isoenzymes, protein and lipoprotein of *Drosophila melanogaster* larvae, pupae and adults. Comp. Biochem. Physiol. **40B**, 709–713 (1971).

PAPPAS, P. W., RODRICK, G. E., DIEBOLT, J.: Protein and enzyme variation in *Drosophila*. Comp. Biochem. Physiol. **40B**, 1029–1035 (1971).

PARZON, S. D., FOX, A. S.: Purification of xanthine dehydrogenase from *Drosophila melanogaster*. Biochim. Biophys. Acta **92**, 465–471 (1964).

PASTEUR, N., KASTRITSIS, C. D.: Developmental studies in *Drosophila* 1. Acid phosphatases, esterases and other proteins in organs and whole-fly homogenates during development of *D. pseudoobscura*. Develop. Biol. **26**, 525–536 (1971).

PATTERSON, E. K., LANG, H. M.: Hydrolysis of leucineamide, leucylglycine and leucylglycylglycine by peptidases. Federation Proc. **13**, 272 (1954).

PEEPLES, E. E., BARNETT, D. R., OLIVER, C. P.: Phenol oxidases of a *lozenge* mutant of *Drosophila*. Science **159**, 548–552 (1968).

PEEPLES, E. E., GEISLER, A., WHITCRAFT, C. J., OLIVER, C. P.: Comparative studies of phenol oxidase activity during pupal development of three *lozenge* mutants *(lz^s, lz, lz^k)* of *Drosophila melanogaster*. Genetics **62**, 161–170 (1969a).

PEEPLES, E. E., GEISLER, A., WHITCRAFT, C. J., OLIVER, C. P.: Activity of phenol oxidases at the puparium formation stage in development of nineteen *lozenge* mutants of *Drosophila melanogaster*. Biochem. Genet. **3**, 563–569 (1969b).

PELLEY, J. W., STAFFORD, D. W.: Partial purification of the aminoacyl-tRNA binding enzyme from *Drosophila* larvae. Biochim. Biophys. Acta **204**, 400–405 (1970a).

PELLEY, J. W., STAFFORD, D. W.: Studies on the enzymatic binding of aminoacyl transfer ribonucleic acid to ribosomes in a *Drosophila in vitro* system. Biochemistry **9**, 3408–3414 (1970b).

PHILLIPS, J. R., FORREST, H. S.: Terminal synthesis of xanthommatin in *Drosophila melanogaster*. II. Enzymatic formation of the phenoxazinone nucleus. Biochem. Genet. **4**, 489–498 (1970).

PHILLIPS, J. R., FORREST, H. S.: Deoxyribonucleic Acid-dependent ribonucleic acid polymerase from *Drosophila melanogaster* embryos. J. Biol. Chem. **248**, 265–269 (1973).

PHILLIPS, J. R., FORREST, H. S., KULKARNI, A. D.: Terminal synthesis of xanthommatin in *Drosophila melanogaster*. III. Mutational pleiotrophy and pigment granule association of phenoxazinase synthetase. Genetics **73**, 45–56 (1973).

PHILLIPS, J. R., SIMMONS, J. R., BOWMAN, J. T.: Re-evaluation of a system for the *in vitro* synthesis of tryptophan pyrrolase. Biochem. Biophys. Res. Commun. **29**, 253–257 (1967).

PHILLIPS, J. R., SIMMONS, J. R., BOWMANN, J. T.: Terminal synthesis of xanthommatin in *Drosophila melanogaster*. I. Roles of phenol oxidase and substrate availability. Biochem. Genet. **4**, 481–487 (1970).

PIPKIN, S. B.: Genetics of octanol dehydrogenase in *Drosophila metzii*. Genetics **60**, 81–92 (1968).

PIPKIN, S. B.: Genetic evidence for a tetramer structure of octanol dehydrogenase of *Drosophila*. Genetics **63**, 405–418 (1969).

PIPKIN, S. B.: BREMNER, T. A.: Aberrant octanol dehydrogenase isozyme patterns in interspecific *Drosophila* hybrids. J. Exp. Zool. **175**, 283–296 (1970).

POSTLETHWAIT, J. H., SCHNEIDERMAN, H. A.: Developmental genetics of *Drosophila* imaginal discs. Ann. Rev. Genet. **1**, 381–433 (1973).

POULSON, D. F., BOELL, E. J.: A comparative study of cholinesterase activity in normal and genetically deficient strains of *Drosophila melanogaster*. Biol. Bull. **31**, 228 (1946).

PRAKASH, S., LEWONTIN, R. C.: A molecular approach to the study of genic heterozygosity in natural populations. III. Direct evidence of coadaptation in gene arrangements of *Drosophila*. Proc. Nat. Acad. Sci. U.S. **59**, 398–405 (1968).

PRAKASH, S., LEWONTIN, C., HUBBY, J. L.: A molecular approach to the study of genic heterozygosity in natural populations. IV. Patterns of genic variation in central, marginal and isolated populations of *Drosophila pseudoobscura*. Genetics **61**, 841–858 (1969).

PRAKASH, S., MERRITT, R. B.: Direct evidence of genic differentiation between sex ratio and standard gene arrangements of chromosome in *Drosophila pseudoobscura*. Genetics **72**, 164–175 (1972).

PRYOR, M. G. M.: Sclerotization. In: FLORKIN, M., MASON, H. S. (Eds.): Comparative biochemistry Vol. IV, pp. 371–396. New York: Academic Press 1962.

RECHSTEINER, M. C.: *Drosophila* lactate dehydrogenase: Partial purification and characterization. J. Insect. Physiol. **16**, 957–977 (1970 a).

RECHSTEINER, M. C.: *Drosophila* lactate dehydrogenase and α-glycerol-phosphate dehydrogenase: Distribution and change in activity during development. J. Insect. Physiol. **16**, 1179–1192 (1970 b).

RICHARDSON, R. H., JOHNSON, F. M.: Isozyme variability in species of the genus *Drosophila*. II. A multiple allelic isozyme system in *Drosophila busckii*: A stable polymorphic system. Biochem. Genet. **1**, 73–79 (1967).

RICHMOND, R. C.: Enzyme variability in the *Drosophila willistoni* group. III. Amounts of variability in the superspecies *D. paulistorum*. Genetics **70**, 87–112 (1972).

RITOSSA, F. M., ATWOOD, K. C., SPIEGELMAN, S.: On the redundancy of DNA complementary to amino acid transfer RNA and its absence from the nucleolar organizer region of *Drosophila melanogaster*. Genetics **54**, 663–676 (1966).

RIZKI, M. T. M.: Intracellular localization of kynurenine in the fat body of *Drosophila*. J. Biophys. Biochem. Cytol. **9**, 567–572 (1961).

RIZKI, M. T. M.: Genetic control of cytodifferentiation. J. Cell Biol. **16**, 513–520 (1963).

RIZKI, M. T. M., RIZKI, R. M.: Functional significance of the crystal cells in the larva of *Drosophila melanogaster*. J. Biophys. Biochem. Cytol. **5**, 235–240 (1959).

RIZKI, M. T. M., RIZKI, R. M.: An inducible enzyme system in the larval cells of *Drosophila melanogaster*. J. Cell Biol. **17**, 87–92 (1963).

SACKTOR, B.: Energetics and respiratory metabolism of muscular contraction. In: ROCKSTEIN, M. (Ed.): Physiology of insecta, Vol. II, pp. 483–580. New York: Academic Press 1965.

SACKTOR, B.: Regulation of intermediary metabolism, with special reference to the control mechanisms in insect flight muscle. Advan. Insect. Physiol. **7**, 267–347 (1970).

SAKAI, R. K., TUNG, D. A., SCANDALIOS, J. G.: Developmental genetic studies of amino peptidases in *Drosophila melanogaster*. Mol. Gen. Genet. **105**, 24–29 (1969).

SAMIS, H. V., BAIRD, M. B., MASSIE, H. R.: Renewal of catalase activity in *Drosophila* following treatment with 3-amino-1,2,4-triazole. J. Insect Physiol. **18**, 991–1000 (1972).

SAYLES, C. D., BROWDER, L. M., WILLIAMSON, J. H.: Expression of xanthine dehydrogenase activity during embryonic development of *Drosophila melanogaster*. Develop. Biol. **33**, 213–217 (1973).

SCHALET, A.: Temperature sensitivity of complementation at the *maroon-like* eye color locus in *Drosophila melanogaster*. Mol. Gen. Genet. **110**, 82–85 (1971).

SCHALET, A., KERNAGHAN, R. P., CHOVNICK, A.: Structural and phenotypic definition of the *rosy* cistron in *Drosophila melanogaster*. Genetics **50**, 1261–1268 (1964).

SCHIMKE, R. T., DOYLE, D.: Control of enzyme levels in animal tissues. Ann. Rev. Biochem. **39**, 929–976 (1970).

SCHNEIDERMAN, H.: Alkaline phosphatase relationships in *Drosophila*. Nature **216**, 604–605 (1967).

SCHNEIDERMAN, H., YOUNG, W. J., CHILDS, B.: Patterns of alkaline phosphatase in developing *Drosophila*. Science **151**, 461–463 (1966).

SEECOF, R. L., KAPLAN, W. D., FUTCH, D. G.: Dosage compensation for enzyme activities in *Drosophila melanogaster*. Proc. Nat. Acad. Sci. U.S. **62**, 528–535 (1969).

SEYBOLD, W. D.: Purification of *Drosophila* xanthine dehydrogenase. Experientia **29**, 758 (1973a).

SEYBOLD, W. D.: Purification and partial characterization of xanthine dehydrogenase from *Drosophila melanogaster*. Biochem. Biophys. Acta **334**, 266 (1973b).

SHAPARD, P. B.: A physiological study of the vermilion eye color mutants of *Drosophila melanogaster*. Genetics **45**, 359–376 (1960).

SHAW, C. R.: Electrophoretic variation in enzymes. Science **149**, 930–943 (1965).

SHAW, C. R., PRASAD, R.: Starch gel electrophoresis of enzymes — A compilation of recipes. Biochem. Genet. **4**, 297–320 (1970).

SHELTON, E. E., SIMMONS, J. R., BOWMAN, J. T.: The enzymatic basis of the "starvation effect" in *vermillion* strains of *Drosophila*. Genetics **56**, 589 (1967).

SHINODA, T., GLASSMAN, E.: Multiple molecular forms of xanthine dehydrogenase and related enzymes. II. Conversion of one form of xanthine dehydrogenase to another by extracts of *Drosophila melanogaster*. Biochim. Biophys. Acta **160**, 178–187 (1968).

SHOUP, J. R.: The development of pigment granules in the eyes of wild type and mutant *Drosophila melanogaster*. J. Cell Biol. **29**, 223–249 (1966).

SIEBER, F., FOX, D. J., URSPRUNG, H.: Properties of octanol dehydrogenase from *Drosophila*. FEBS Letters **26**, 274–276 (1972).

SMITH, K. D., URSPRUNG, H., WRIGHT, T. R. F.: Xanthine dehydrogenase in *Drosophila*: Detection of isozymes. Science **142**, 226–227 (1963).

SMITH, P. D., FINNERTY, V. G., CHOVNICK, A.: Gene conversion in *Drosophila*: Non-reciprocal events at the *maroon-like* cistron. Nature **228**, 441–444 (1970).

SOFER, W. H., HATKOFF, M. A.: Chemical selection of alcohol dehydrogenase negative mutants in *Drosophila*. Genetics **72**, 545–549 (1972).

SOFER, W. H., URSPRUNG, H.: *Drosophila* alcohol dehydrogenase: Purification and partial characterization. J. Biol. Chem. **243**, 3110–3115 (1968).

STEELE, M. W., YOUNG, W. J., CHILDS, B.: Genetic regulation of glucose 6-posphate dehydrogenase activity in *Drosophila melanogaster*. Biochem. Genet. **3**, 359–370 (1969).

STEFFENSEN, D. M., WIMBER, D. E.: Localization of tRNA genes in the salivary chromosomes of Drosophila by RNA: DNA hybridization. Genetics **69**, 163–178 (1971).

STEFFENSEN, D. M., WIMBER, D. E.: Hybridization of nucleic acids to chromosomes. In: URSPRUNG, H. (Ed.): Nucleic acid hybridization in the study of cell differentiation. Results and problems in cell differentiation, Vol. 3, pp. 47–63. Berlin-Heidelberg-New York: Springer 1972.

STEWART, B. R., MERRIAM, J. R.: Segmental aneuploidy and enzyme activity as a method for cytogenetic localisation in *Drosophila melanogaster*. Genetics **76**, 301–309 (1974).

STONE, W. S., WHEELER, M. R., JOHNSON, F. M., KOJIMA, K.: Genetic variation in natural island populations of members of the *Drosophila nasuta* and *D. ananassae* subgroups. Proc. Nat. Acad. Sci. U.S. **59**, 102–109 (1968).

SULLIVAN, D. T., GRILLO, S. L., KITOS, R. J.: Subcellular localization of the first three enzymes of the ommochrome synthetic pathway in *Drosophila melanogaster*. J. Exp. Zool. **188**, 225–234 (1974).

SULLIVAN, D. T., KITOS, R. J., SULLIVAN, M. C.: Developmental and genetic studies on kynurenine hydroxylase from *Drosophila melanogaster*. Genetics 75, 651–661 (1973).

SULLIVAN, D. T., MITCHELL, H. K.: The role of the S component in the activation of phenol oxidase from *Drosophila melanogaster*. Federation Proc. **28**, 900 (1969).

SUZUKI, D. T.: Temperature-sensitive mutations in Drosophila melanogaster. Science **170**, 695–706 (1970).

SUZUKI, D. T., PITERNICK, L. K., HAYASHI, S., TARASOFF, M., BAILLIE, D., ERASMUS, U.: Temperature-sensitive mutations in *Drosophila melanogaster*. I. Relative frequencies among X-ray and chemically induced sex-linked recessive lethals and semilethals. Proc. Nat. Acad. Sci. U.S. **58**, 907–912 (1967).

TARTOF, K. D.: Interacting gene systems: I. The regulation of tryptophan pyrrolase by the *vermilion-suppressor* of *vermilion* system in *Drosophila*. Genetics **62**, 781–795 (1969).

TATUM, E. L., BEADLE, G. W.: Effect of diet on eye color development in *Drosophila melanogaster*. Biol. Bull. 77, 415–422 (1939).

TOBARI, Y. N., KOJIMA, K.: A study of spontaneous mutation rates at ten loci detectable by starch gel electrophoresis in *Drosophila melanogaster*. Genetics **70**, 397–403 (1972).

TOBLER, J., BOWMAN, J. T., SIMMONS, J. R.: Gene modulation in *Drosophila*: Dosage compensation and relocated v^+ genes. Biochem. Genet. **5**, 111–117 (1971).

TRIANTAPHYLLIDIS, C. D.: Allozyme variation in populations of *Drosophila melanogaster* and *D. simulans* from Northern Greece. J. Heredity **64**, 69–72 (1973).

TRIPPA, G., SANTOLAMAZZA, C., SCOZZARI, R.: *Phosphoglucomutase (Pgm)* locus in *Drosophila melanogaster*: Linkage and population data. Biochem. Genet. **4**, 665–667 (1970).

TUNNICLIFF, G., RICK, J. T., CONNOLLY, K.: Locomotor activity in *Drosophila*. V. A comparative biochemical study of selectively bred populations. Comp. Biochem. Physiol. **29**, 1239–1245 (1969).

TWARDZIK, D. R., GRELL, E. H., JACOBSON, K. B.: Mechanism of suppression in *Drosophila*: A change in tyrosine transfer RNA. J. Mol. Biol. **57**, 231–245 (1971).

UDA, F., MATSUMIYA, H., TAIRA, T.: Enzymatic deamination of adenosine 2′, 3′-cyclic phosphate in *Drosophila melanogaster*. Biochem. Biophys. Res. Commun. **34**, 472–479 (1969).

URSPRUNG, H.: Weitere Untersuchungen zu Komplementarität und Nicht-Autonomie der Augenfarbmutanten *ma-l* and *ma-l*^{bz} von *Drosophila melanogaster*. Z. Vererbungslehre **92**, 119–125 (1961).

URSPRUNG, H.: XIII. International Congress of Genetics. Genetics, in press (1974).

URSPRUNG, H., CARLIN, L.: *Drosophila* alcohol dehydrogenase: *In vitro* changes of isozyme patterns. Ann. N.Y. Acad. Sci. **151**, 456–475 (1968).

URSPRUNG, H., CONSCIENCE-EGLI, M., FOX, D. J., WALLIMAN, T.: Origin of leg musculative during *Drosophila* metamorphosis. Proc. Nat. Acad. Sci. U.S. **69**, 2812–2813 (1972).

URSPRUNG, H., HADORN, E.: Xanthindehydrogenase in Organen von *Drosophila melanogaster*. Experientia **17**, 230–232 (1961).

URSPRUNG, H., LEONE, J.: Alcohol dehydrogenases: A polymorphism in *Drosophila melanogaster*. J. Exp. Zool. **160**, 147–154 (1965).

URSPRUNG, H., SMITH, K. D., SOFER, W. H., SULLIVAN, D. T.: Assay systems for the study of gene function. Science **160**, 1075–1081 (1968).

URSPRUNG, H., SOFER, W. H., BURROUGHS, N.: Ontogeny and tissue distribution of alcohol dehydrogenase in *Drosophila melanogaster*. Wilhelm Roux' Arch. **164**, 201–208 (1970).

VYSE, E. R., NASH, D.: Nutritional conditional mutants of *Drosophila melanogaster*. Genet. Res. (Camb.) **13**, 281–287 (1969).

VYSE, E. R., SANG, J. H.: A purine and pyrimidine requiring mutant of *Drosophila melanogaster*. Genet. Res. (Camb.) **18**, 117–121 (1971).

WADDINGTON, C. H.: Body color genes in *Drosophila*. Proc. Zool. Soc. (Lond.), Ser. A **111**, 173–180 (1941).

WALDNER-STIFELMEIER, R. D.: Untersuchungen über die Proteasen im Wildtyp und in den Letalmutanten (*lme* and *ltr*) von *Drosophila melanogaster*. Z. Vergleich. Physiol. **56**, 268–269 (1967).

WALLIS, B. B., FOX, A. S.: Genetic and developmental relationships between two alkaline phosphatases in *Drosophila melanogaster*. Biochem. Genet. **2**, 141–158 (1968).

WARD, C. L., BIRD, M. A.: Comparative studies of cytochrome c oxidase activity and mutability in two strains of *Drosophila*. Genetics **47**, 99–107 (1962).

WARD, C. L., BIRD, M. A.: Cytochrome oxidase activity in chromosome interchange stocks of the Oslo and Iso-Amherst strains of *Drosophila melanogaster*. Genetics **48**, 1435–1440 (1963).

WEISS, H.: Studies on mutations affecting aldehyde oxidase in *Drosophila*. Thesis, Reed College 1972.

WESSING, A., EICHELBERG, D.: Die fluoreszierenden Stoffe aus den malpighisschen Gefäßen der Wildform und verschiedener Augenfarbenmutanten von *Drosophila melanogaster*. Z. Naturforsch. **23b**, 376–386 (1968).

WHIPPLE, H.E. (Ed.): Gel electrophoresis. Ann. N.Y. Acad. Sci. **121**, 305–650 (1964).

WHITMORE, D., JR., WHITMORE, E., GILBERT, L.I.: Juvenile hormone induction of esterases: A mechanism for the regulation of juvenile hormone titer. Proc. Nat. Acad. Sci. U.S. **69**, 1592–1595 (1972).

WHITNEY, J.B., LUCCHESI, J.C.: Ontogenetic expression of fumarase activity in *Drosophila melanogaster*. Insect Biochem. **2**, 367–370 (1972).

WRIGHT, D.A., SHAW, C.R.: Genetics and ontogeny of α-glycerophosphate dehydrogenase isozymes in *Drosophila melanogaster*. Biochem. Genet. **3**, 343–353 (1969).

WRIGHT, D.A., SHAW, C.R.: Time of expression of genes controlling specific enzymes in *Drosophila* embryos. Biochem. Genet. **4**, 385–394 (1970).

WRIGHT, T.R.F.: The genetic control of an esterase in *Drosophila melanogaster*. Am. Zool. **1**, 476 (1961).

WRIGHT, T.R.F.: The genetics of an esterase in *Drosophila melanogaster*. Genetics **48**, 787–801 (1963).

WRIGHT, T.R.F.: The phenogenetics of temperature sensitive alleles of *lethal myospheroid* in *Drosophila*. Proc. 12th Intern. Congr. Genet., Tokyo 1968, Vol. 1, p. 141.

WRIGHT, T.R.F.: The genetics of embryogenesis in *Drosophila*. Advan. Genet. **15**, 262–395 (1970).

WRIGHT, T.R.F., MACINTYRE, R.J.: A homologous gene-enzyme system, esterase 6, in *Drosophila melanogaster* and *D. simulans*. Genetics **48**, 1717–1726 (1963).

WRIGHT, T.R.F., MACINTYRE, R.J.: Heat-stable and heat-liable Esterase-6^F enzymes in *Drosophila melanogaster* produced by different *Est.*6^F *alleles. J. Elisha Mitchell Sci. Soc.* **81**, 17–19 (1965).

YAMAZAKI, H.I.: The cuticular phenol oxidase in *Drosophila virilis*. J. Insect Physiol. **15**, 2203–2211 (1969).

YAMAZAKI, H.I., OHNISHI, E.: Phenol oxidase activity and phenotypic expression of the melanotic tumur strain tu^g in *Drosophila melanogaster*. Genetics **59**, 237–243 (1968).

YAMAZAKI, T.: Measurement of fitness at the *esterase-5* locus in *Drosophila pseudoobscura*. Genetics **67**, 579–603 (1970).

YAO, T.: Cytochemical studies on the embryonic development of *Drosophila melanogaster*. II. Alkaline and acid phosphatases. Quart. J. Microscop. Sci. **91**, 79–88 (1950).

YARBROUGH, K., KOJIMA, K.: The mode of selection at the polymorphic *esterase-6* locus in cage populations of *Drosophila melanogaster*. Genetics **57**, 677–681 (1967).

YEN, T.T., GLASSMAN, E.: Electrophoretic variants of xanthine dehydrogenase in *Drosophila melanogaster*. Genetics **52**, 977–981 (1965).

YEN, T.T., GLASSMAN, E.: Electrophoretic variants of xanthine dehydrogenase in *Drosophila melanogaster*. Biochim. Biophys. Acta **146**, 35–44 (1967).

YOUNG, W.J.: X-linked electrophoretic variation in 6-phosphogluconate dehydrogenase in *Drosophila melanogaster*. J. Heredity **57**, 58–60 (1966).

YOUNG, W.J., PORTER, J.E., CHILDS, B.: Glucose-6-phosphate dehydrogenase in *Drosophila:* X-linked electrophoretic variants. Science **143**, 140–141 (1964).

ZIEGLER, I.: Genetic aspects of ommochrome and pterin pigments. Advan. Genet. **10**, 349–403 (1961).

ZIEGLER, I., HARMSEN, R.: The biology of pteridines in insects. Advan. Insect Physiol. **6**, 140–203 (1969).

ZOUROS, E., KRIMBAS, C.B.: Evidence for linkage disequilibrium maintained by selection in two natural populations of *Drosophila subobscura*. Genetics **73**, 659–674 (1973).

ZWEIDLER, A., COHEN, L.H.: Large scale isolation and fractionation of organs of *Drosophila melanogaster* larvae. J. Cell. Biol. **51**, 240–248 (1971).

Subject Index

Results and Problems in Cell Differentiation

A Series of Topical Volumes in Developmental Biology

Editors:
W. Beermann, Tübingen
J. Reinert, Berlin
H. Ursprung, Zürich

Vol. 1
The Stability of the Differentiated State

Edited by H. Ursprung
56 figs. XI, 154 pp. 1968
Cloth DM 59,–; US $24.10
ISBN 3-540-04315-2

This volume brings together authoritative reviews on the subject from the point of view of tissue culture (both in vitro and vivo), regeneration, dedifferentiation, metaplasia, clonal analysis, and somatic hybridization. The ten contributing authors are actively working in the respective areas of interest and have written short, concise accounts of their own and related work. The book is intended for advanced students who want to keep up with recent developments in these areas of basic biology which are so important for the understanding of the development of higher organisms, both normal and malignant. The book as a whole uncovers similarities in experimental systems as widley different as plants and insects and provides new insights into fundamental processes.

Vol. 2
Origin and Continuity of Cell Organelles

Edited by J. Reinert and H. Ursprung
135 fig. XIII, 342 pp. 1971
Cloth DM 72,–; US $29.40
ISBN 3-540-05239-9

A great deal of precise information is available on structure and function of cell organelles. But much controversy exists on the question of their origin, and the mode of their replication. This book contains authoritative reviews relevant to this question on the following components of cells: mitochondria, vacuoles, desmosomes, centrioles, polar granules, Golgi apparatus, endosymbionts, plastids, microtubules, lysosomes, and membranes in general.

Vol. 3
Nucleic Acid Hybridization in the Study of Cell Differentiation

Edited by H. Ursprung
29 fig. XI, 76 pp. 1972
Cloth DM 36,–; US $14.70
ISBN 3-540-05742-0

The method of molecular hybridization of nucleic acids has made it possible to obtain more precise information on quantitative and – in some cases at least – qualitative composition of the genome. Both beginners and advanced students will find the book helpful for understanding the principles of the method, the various techniques that are used, and the biological implications of the results already obtained.

Vol. 4
Developmental Studies on Giant Chromosomes

Edited by W. Beermann
110 fig. XV, 227 pages
1972. Cloth DM 59,–;
US $24.10
ISBN 3-540-05748-X

This volume is devoted to the question of structural and functional organization of the eucaryotic genome in relation to cell differentiation as revealed by current work on the genetic fine structure and synthetic activities of giant polytene chromosomes and their subunits, the chromomeres.

Vol. 5
The Biology of Imaginal Disks

Edited by H. Ursprung and R. Nöthiger
56 fig. XVII, 172 pp. 1972
Cloth DM 46,–; US $18.80
ISBN 3-540-05785-4

The book summarizes the current state of knowledge concerning the biochemistry and ultrastructure of the imaginal disk. It reviews experiments of the mechanisms of cell determination as studied in imaginal disks.

Prices are subject to change without notice

Springer-Verlag
Berlin
Heidelberg
New York